W9-CDW-576

Withdrawn

JAN 2 8 2005

MAR 0 7 2005

DEC 2 9 2009

Look for these other CraftMaster books

And these other McGraw-Hill books

WELDING

Don Geary

McGraw-Hill

New York   San Francisco   Washington, D.C.   Auckland   Bogotá
Caracas   Lisbon   London   Madrid   Mexico City   Milan
Montreal   New Delhi   San Juan   Singapore
Sydney   Tokyo   Toronto

Library of Congress Cataloging-in-Publication Data

Geary, Don.
  Welding / Don Geary.
      p.   cm.
  ISBN 0-07-134245-1
  1. Oxyacetylene welding and cutting.   I. Title.
TS228.G425   2000
671.5'2 — dc21   99-35501
                                                                CIP

McGraw-Hill

*A Division of The **McGraw·Hill** Companies*

Copyright © 2000 by The McGraw-Hill Companies, Inc. All rights reserved. Printed in the United States of America. Except as permitted under the United States Copyright Act of 1976, no part of this publication may be reproduced or distributed in any form or by any means, or stored in a data base or retrieval system, without the prior written permission of the publisher.

    6 7 8 9 0   FGR/FGR   0 9 8 7 6 5 4

ISBN 0-07-134245-1

The sponsoring editor for this book was Zoe Foundatos, the editing supervisor was Peggy Lamb, and the production supervisor was Pamela Pelton. It was set in Melior per the CMS design by Joanne Morbit of McGraw-Hill's Professional Book Group composition unit in Hightstown, N.J.

Printed and bound by Quebecor / Fairfield

Previously published under the title *The Welder's Bible.*

 This book is printed on recycled, acid-free paper containing a minimum of 50% recycled de-inked fiber.

McGraw-Hill books are available at special quality discounts to use as premiums and sales promotions, or for use in corporate training programs. For more information, please write to the Director of Special Sales, McGraw-Hill, Professional Publishing, Two Penn Plaza, New York, NY 10121-2298. Or contact your local bookstore.

Information contained in this work has been obtained by The McGraw-Hill Companies, Inc. ("McGraw-Hill") from sources believed to be reliable. However, neither McGraw-Hill nor its authors guarantee the accuracy or completeness of any information published herein, and neither McGraw-Hill nor its authors shall be responsible for any errors, omissions, or damages arising out of use of this information. This work is published with the understanding that McGraw-Hill and its authors are supplying information but are not attempting to render engineering or other professional services. If such services are required, the assistance of an appropriate professional should be sought

Northville District Library
212 W. Cady Street
Northville, MI 48167-1560   JAN 2 0 2005   3 9082 09521 0103

CONTENTS

PREFACE

In this book you will learn the basics of the oxyacetylene process for joining and cutting metals. Oxyacetylene welding and cutting have been in use for over 100 years and, this basic method is still in widespread use today all over the world. I should mention, at this point, that I do not advise the do-it-yourselfer to teach himself or herself how to weld with any method other than oxyacetylene. While the text is designed for the do-it-yourself welder, there are many descriptions of professional welding and cutting techniques. I trust you will find these additions helpful in learning how to weld. If any terms are unfamiliar to you, check the glossary. I have tried to make it as complete and helpful as possible.

I will caution you in the beginning, as well as throughout the book, that working with oxyacetylene welding or cutting equipment can be dangerous. You, as a welder, must always approach any welding problem soberly. Use your best judgment at all times. Safe welding and cutting should always be your first priority. In this way you will develop safe and sound welding habits that can result in a long and healthy welding experience. Safety is of such a concern to me that I have devoted an entire chapter to pointing out how to safely handle welding equipment, as well as how to deal with mishaps such as fire around your welding shop. Let me suggest that you read this chapter before setting up your welding equipment or lighting your torch for the first time.

While it is entirely possible to become a proficient welder solely with the aid of this book and many hours of practice, you should supplement the information contained in these pages with advice and instruction from a professional in the art of welding. Spend the time to seek out professional welders in your area. These people are often warehouses of both practical and technical information. If time permits, you should also enroll in some type of educational program that deals specifically with welding and cutting—adult education programs, offered by continuing-education schools in your area, for example. Often the cost is minimal, and you will get the opportunity to learn some of the things that cannot possibly be found in any book about welding.

If you do not at this time own an oxyacetylene welding or cutting torch but are considering buying one, buy the best you can afford. I suggest that you pay a visit to a welding supply house in your area. Look in the Yellow Pages and ask for advice. More often than not, the people who work at supplying welders with tools and materials are very helpful to beginners. Presently, a basic welding and cutting outfit can be had for about $200. Used equipment, in good condition, may cost half as much. As far as I am concerned, the only way to determine if used welding equipment is in good condition is to have it checked over by a professional who specializes in welding equipment.

It will take you many hours to learn how to join or cut metal with the oxyacetylene process. Devote the time to learn how to do things properly. In the long run this is always the best approach.

SAFE WELDING - Don Geary

WELDING

Oxyacetylene Welding Fuels

Welding is a relatively new way of joining two or more pieces of metal in such a way as to make the finished piece as strong as the original metal. The oldest type of welding, *oxyacetylene welding,* was developed about 100 years ago. There have been many developments in metal joining processes since then.

For our purposes, metal joining can be broken down into three rather broad categories—*gas welding, electric welding,* and *gas-electric welding.* Since the home welder is not interested in production work, there is little need, other than for informational purposes, to learn about other types of welding such as *laser* or *plasma welding.* These processes are used solely in industry.

History of Gas Welding

Gas welding, the oldest and easiest of all types of welding to learn, is very simple in principle. Basically, oxygen and acetylene are burned together to produce a flame that is hotter than the melting point of most metals. The temperature of an oxyacetylene flame is generally accepted at being around 6,000°F.

In view of the fact that oxyacetylene is so widely used, it is almost unbelievable that this process did not come into existence until the

1

beginning of this century. Oxyacetylene welding was first made possible through the experiments and discoveries of a French chemist, Henri Le Chatelier, in 1895. He was the first to discover that burning oxygen and acetylene produced a flame with a temperature far higher than that of any other flame in existence.

It wasn't long before the capabilities of oxyacetylene became known to the industrial world and were put to use. After a workable way to store and transport oxygen and acetylene was developed, the road was clear for widespread use of this new method of joining metals.

It was during World War I that the use of oxyacetylene welding accelerated. The pressures to supply a fighting army and to repair existing heavy equipment brought worldwide attention to oxyacetylene welding. After the war, there arose a need for greater controls over the welding process, and machines were developed that could weld.

The oxyacetylene welding-cutting process is the most versatile means of working with metals. No other equipment or process in use by the metal industries is capable of performing such a wide variety of work on most types and thicknesses of metals. The oxyacetylene method of welding is also the easiest to master and probably the most versatile for the do-it-yourself welder. The bulk of this book will be concerned with this type of welding.

Actually the oxyacetylene process can be used for *joining, heating,* and *cutting* metals. Joining, or fusion, welding is an important application of the oxyacetylene process. Here the two edges of a metal are heated up to their melting points and fused together (Fig. 1.1).

Heating with the oxyacetylene process is often used for forming metals into various shapes and heat treating metals in operations such as annealing, flame hardening, tempering, case hardening, and stress relieving (Fig. 1.2).

The last important use of oxyacetylene is to cut metals. To do so, a stream of pure oxygen is directed against an area of heated metal. This action causes the metal to oxidize, or burn, and thus be cut (Fig. 1.3).

It is obvious from the name "oxyacetylene" that this type of welding uses a combination of oxygen and acetylene. To understand these basic components, it may be helpful to discuss these two substances and explain their significance to the welding process.

FIGURE 1.1

Welding is a metalworking process in which the joint metal is heated up to the melting point and the pieces are fused together.

Oxygen

Oxygen is present in the air we breathe but in small amounts. About one-fifth of our atmosphere is pure oxygen. By contrast, oxygen used in the welding process is about as pure as possible—over 99 percent pure. The method of producing pure oxygen for welding and medical purposes is called the *liquid-air process.*

Liquid-Air Process

The following is an oversimplification, but it is an adequate description of the liquid-air process. Atmospheric air, as mentioned earlier, consists of about 20 percent pure oxygen, 78 percent nitrogen, and 2 percent other gases (by volume). Oxygen and nitrogen have different boiling temperatures. Thus, using the liquid-air process, the two can be separated by heating atmospheric air to a certain temperature and

FIGURE 1.2

Welding is often used to heat and shape metals, as is being done to this automobile fender.

holding it at this temperature until the nitrogen, which has a boiling point of 295°F, boils off. After the nitrogen has been removed from atmospheric air, oxygen and a small amount of other gases remain. These include carbon dioxide, argon, hydrogen, neon, and helium. Oxygen has the highest boiling point of all these gases; thus to separate it completely, the remaining mixture is further heated until only pure oxygen remains. The pure oxygen is then stored as either a gas or liquid, depending on its eventual use. The liquid-air process is probably the most widely used method of producing pure oxygen.

Cylinders

Oxygen is commonly sold in cylinders in three sizes: 244 cubic feet, 122 cubic feet, and 80 cubic feet (Fig. 1.4). There are very strict requirements for oxygen cylinders. They must be able to withstand over 1 ton of pressure per square inch. The Interstate Commerce

FIGURE 1.3

Cutting can also be accomplished with welding equipment.

Commission (ICC) has set up guidelines for oxygen cylinders. No part of the cylinder may be less than ¼ inch thick. Each cylinder must be made or forged from a single piece of steel. The steel itself must be armor plate, high-carbon steel (Fig. 1.5).

Since the ICC requires periodic inspection of oxygen cylinders, which are shipping containers, very few oxygen cylinders are actually owned by private individuals. The oxygen supply houses, which own and lease oxygen cylinders, are responsible for complying with ICC regulations and inspections. This makes the whole affair easier on the consumer.

FIGURE 1.4

The three most popular sizes of portable oxygen cylinders, left to right, are 244 cubic feet, 122 cubic feet, and 80 cubic feet.

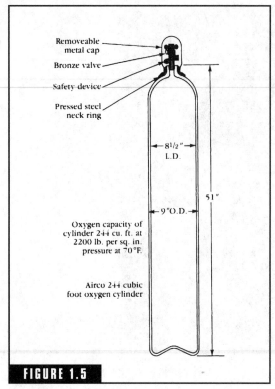

Removeable metal cap

Bronze valve

Safety device

Pressed steel neck ring

8½" I.D.

51"

9" O.D.

Oxygen capacity of cylinder 244 cu. ft. at 2200 lb. per sq. in. pressure at 70°F.

Airco 244 cubic foot oxygen cylinder

FIGURE 1.5

A typical oxygen cylinder.

The universally accepted color for oxygen cylinders, lines, and control knobs is green. However, there is no regulation that requires oxygen cylinders to be green in color; thus, many companies paint their oxygen cylinders a special identifying color. It will be to your advantage to become familiar with the color used by your dealer. Often two oxygen supply houses in the same city will have two different colors for their oxygen tanks.

Safety Practices

Oxygen cylinders are not dangerous when used and stored according to generally accepted safety precautions. Nevertheless, some general comments are in order.

Use only cylinders carrying ICC markings. You can be certain that cylinders of this type comply with the stringent regulations of the ICC.

Store cylinders only in a safe location, and fasten them in place. This will ensure that the cylinder cannot be knocked over. Keep tanks in an area away from stoves, radiators, furnaces, or other overly warm places. Oxygen cylinders should also be kept away from all combustible materials or liquids. If cylinders are stored in the open, they should be protected from the elements of water, heat, cold, and the sun's direct rays.

Never use oxygen cylinders for any purpose other than holding oxygen. You should never use oxygen cylinders as rollers to help move large or heavy objects, for example. Nor should you use oxygen cylinders as supports.

Oxygen cylinders should be stored and used in an upright position. In most cases, a hand truck specifically designed to hold two cylinders (both oxygen and acetylene) is the best means for storing and using oxygen. The cylinders should be securely chained or strapped to the cart to prevent them from ever falling over and allowing them to be transported safely (Fig. 1.6).

Never use the valve on top of the cylinder to lift the cylinder from a horizontal to a vertical position. The best way to lift a cylinder is to first make sure that the valve protection cap is secured tightly. Then raise the cylinder by grasping the cap firmly and lifting (Fig. 1.7).

Never allow oxygen cylinders to come in contact with live electrical wires or other electrical equipment. Keep cylinders away from welding and cutting work. Make certain that the hoses containing oxygen and acetylene do not lie under the work.

Open cylinder valves all the way when working, and always close cylinder valves when you have completed working. Never leave the cylinder valve open when you are not in the immediate vicinity. Cylinder valves should always be tightly closed when not in use—whether they are empty or full. If you stop working for lunch, for example, turn off the valves and bleed the lines. I will explain this procedure later (Fig. 1.8).

Never use oxygen around oil or grease as these substances burn violently in the presence of pure oxygen under pressure. Do not use any oil or grease on the regulator fitting.

Never use oxygen for anything other than welding and cutting. Pure oxygen is flammable. There is a very real danger if it is used to "dust off" work or clothing. Similarly, oxygen should never be used for ventilation, pressure tests of any kind, or for any other purpose.

QUICK»TIP

You can rent oxygen (as well as acetylene) cylinders for a fixed period of time. When you require more oxygen, you simply return the empty cylinder to the dealer, who replaces it with a full one. All you pay for is the oxygen. The oxygen itself is what you use. In most parts of the country a 20-year lease is available and quite reasonable. In my case, the cost of a cylinder for oxygen and a cylinder for acetylene costs about $10 a year. You can buy oxygen and acetylene to fill the tanks as you need them.

FIGURE 1.6

Both oxygen and acetylene tanks should be chained to prevent their accidental falling.

FIGURE 1.7

Whenever you pick up a cylinder of oxygen or acetylene, always lift it by the protective cap.

FIGURE 1.8

Always turn off the flow of both oxygen and acetylene at the tank when you stop work for more than 10 minutes.

Acetylene

Acetylene is produced by combining calcium carbide and water. When calcium carbide, commonly called *carbide* in the trade, is dropped into water, a reaction occurs that causes gas bubbles to rise. The gas is acetylene. It has a peculiar odor, and, if lighted, it burns with a black, smoky flame. After the action of the carbide has ceased, a whitish residue remains in the water. This end product is *hydrated, or slaked, lime (calcium hydroxide)*, and it has many uses in the fertilizer industry.

It is interesting, as a side note, to mention that old-time miners' headlamps are called *carbide lamps,* and they work simply by adding water to carbide pellets. A gas is formed in a special container and is forced out of a jet. This gas is ignited in front of a reflector plate, which throws a bright beam of light (Fig. 1.9).

Carbide-to-Water Generators

Acetylene is produced commercially by basically the same process as used in a miner's headlamp, only on a much larger scale. Companies that use great amounts of acetylene will usually have their own generator that supplies them with all of the acetylene they can use. These generators are called *carbide-to-water generators,* and they are almost totally automated. In these types of generators, small amounts of calcium carbide are fed into a large sealed container of water. The heat given off as the reaction occurs is absorbed by the water. The acetylene gas is captured when it rises to the top of the tank. This gas is then either used directly or stored for future use (Fig. 1.10).

Containers

For the home welder, the only practical way to obtain acetylene is in a special tank, which is similar to but much smaller than a standard oxygen cylinder. The three common sizes of acetylene cylinders are 300 cubic feet, 100 cubic feet, and 60 cubic feet. A full tank of acetylene will register approximately 225 psi (pounds per square inch) on a pressure gauge (Fig. 1.11).

As with oxygen, the ICC has set up guidelines and regulations for acetylene containers. One ICC regulation is that free acetylene may not be stored at pressures over 15 pounds per square inch. It is not a problem to stay within this amount because all welding and cutting with oxyacetylene can be performed at pressures of less than 15 psi.

To store acetylene below the 15-psi requirement, special tanks have been developed. It is interesting to note how acetylene is stored. After a tank has been made of heavy-gauge steel, it is packed with a porous substance such as pith from cornstalks, fuller's earth, lime silica, or other materials. Next, the tank is filled with *acetone,* a liquid chemical having the ability of absorbing about 25 times its own volume of acetylene. When the cylinder is filled to about 225 psi with acetylene, the acetone dissolves the acetylene and makes it safe because the acetylene is not free (Fig. 1.12).

Acetylene cylinders have a valve on top to which the regulator is attached. This valve is similar to the valve on oxygen cylinders except that it has left-handed threads. You must turn the valve counterclockwise to open it and clockwise to close it. This turning action is exactly opposite of a standard control valve.

All acetylene tanks have safety plugs on the top or bottom of the cylinder. These plugs are designed to melt, in the event of a fire, around 212°F. Needless to say, the cylinders should be stored and used

FIGURE 1.9

A miner's carbide lamp works on the same basic principle as does an acetylene generator.

QUICK»TIP

Note that all acetylene connections on the cylinder, hoses, and torch handle are left-handed. Each brass fitting has a groove cut around its circumference for quick identification (Fig. 1.13).

FIGURE 1.10

A cutaway view of an acetylene generator.

away from any heat source. Once these safety plugs melt, all of the acetylene and acetone in the cylinder is allowed to escape (Fig. 1.14).

Precautions

As with oxygen cylinders, there are a few general safety precautions that should always be followed. These include the following guidelines for safe operation and storage of acetylene.

Always call acetylene by its proper name: "acetylene." It should never be called "gas" as it is very different from the type of gas used in a kitchen, furnace, or automobile.

Acetylene should be stored away from all heat sources and other fuels. An ideal storage condition is a cool, dry area away from all combustible materials.

Handle acetylene cylinders with special care. The safety plugs melt at about the same temperature as boiling water, and they can "blow out" as a result of rough handling. As with oxygen, cylinders of acetylene should be secured while being used to eliminate the possibility of the cylinder falling over. A specially designed hand truck is probably the best way of moving and storing acetylene and oxygen.

Do not open the valve on an acetylene cylinder more than 1½ turns as this may cause some of the acetone in the tank to escape with the acetylene. Acetone will damage all rubber and plastic parts in the system including the hose and regulator. Therefore, it should not be allowed to escape.

Always use an acetylene cylinder in an upright position. Never use it when the cylinder is in any position other than vertical.

FIGURE 1.11

The three most popular sizes of portable acetylene cylinders, left to right, are 300 cubic feet, 100 cubic feet, and 60 cubic feet. Cylinders of other shapes are also available.

FIGURE 1.12

A cutaway view of a typical acetylene cylinder showing the filler material.

FIGURE 1.13

A fitting for an acetylene regulator. Notice that the nut has grooves around the middle. All left-handed-threaded fittings have these grooves.

FIGURE 1.14

The safety plug on an acetylene cylinder is specially designed to melt at temperatures in excess of 212° F.

Oxyacetylene Welding Equipment

This chapter will examine the important parts of the oxyacetylene welding system. The chapter includes discussion of the proper use of the equipment, which, of course, results in safe and better heating, welding, and cutting.

Regulators

As you know, oxygen and acetylene are commonly used from special cylinders. A full tank of oxygen will tip the scales at about 2,200 psi, and acetylene will measure at about 225 psi. Most welding and cutting can be done with a maximum of 50 psi for oxygen and 10 psi of acetylene. Obviously, some type of device must be attached to the tanks to reduce the high pressure of the tank to a lower working pressure. In addition, this device must also provide a steady flow of oxygen or acetylene over the life of the tank. The pressure in a full tank is obviously more than in a half-full tank, but the working pressure will be the same. Another requirement of such a device is that it must regulate consistently the flow of oxygen or acetylene from the tank to the torch.

There are basically two different types of regulators for the oxyacetylene welder: single-stage and two-stage. Let's now look at both types.

FIGURE 2.1

An oxygen regulator.

Single-Stage Regulators

A *single-stage regulator,* as the name implies, reduces the pressure coming out of a cylinder to a desired working pressure in one step. These regulators are less expensive than two-stage regulators. They are practical for general welding, cutting, and heating. A single-stage regulator will show more of a pressure rise or drop as the cylinder contents are used than will a two-stage regulator. Generally speaking, this deviation in pressure will not affect the quality of the flame on small, short-duration work. But readjustments of the torch controls will be required when welding for a prolonged period (Fig. 2.1).

Two-Stage Regulators

A *two-stage regulator* will maintain practically constant flow pressures during a long period of operation. Two-stage regulators are more expensive than single-stage regulators, but they offer much more precise flow regulation. Generally speaking, all two-stage regulators have two gauges: One shows the pressure in the tank, and the other shows the working pressure in the line.

Actually the term *two stage* means that the pressure in a tank is reduced in two steps before it is allowed into the line. The first stage serves as a high-pressure reduction chamber, indicating the pressure in the tank. The second stage is, in effect, a low-pressure reduction chamber, and it is adjusted by means of a special control knob on the face of the regulator.

There are many types of regulators being produced in this country. All of them, however, fall into either a single- or two-stage classification. Many of these regulators have two gauges. One gauge indicates the pressure in the cylinder, and the other indicates the working pressure being delivered to the torch (Fig. 2.2). Other regulators will have only one gauge, usually showing the working pressure being supplied to the torch. These single-gauge regulators are most commonly single-stage regulators. There are also regulators that do not have any gauges.

Generally these gaugeless regulators are used in commercial work, and they are attached to a pipeline supplying oxygen and acetylene to a workstation (Fig. 2.3).

FIGURE 2.2

While all gauges are constructed of sturdy materials, they can nevertheless be damaged by carelessness or improper use. Internal working parts commonly consist of gears and springs that operate the dial indicator on the face of the unit. Exercise care when handling, installing, or removing any pressure regulator gauge. In addition, store gauges in a safe, dust-free environment to attain a long useful life.

These two-stage regulators have gauges that show working pressure in the lines (left gauge) and pressure in the tank. The regulator on the left side is for acetylene, and the other is for oxygen.

Differences

Probably the greatest disadvantage of using a single-stage regulator, rather than a two-stage regulator, is that a single-stage regulator will require adjustments to the torch as the pressure in the tank is lessened. One other objection to single-stage regulators is that they have a tendency to freeze in cold weather. A sudden expansion and resulting drop in initial pressure causes the gas to cool very rapidly. The end result is a frozen single-stage regulator.

There is a major difference between the regulator used for oxygen and the one used for acetylene. Acetylene brass fittings are left-handed (and marked) and therefore cannot be used on oxygen tanks. Oxygen regulators are considerably stronger overall because of the pressure of standard oxygen tanks. A full tank will contain approximately 2,200 psi of pressure, while a full acetylene tank will contain only 225 psi.

FIGURE 2.3

A typical acetylene regulator without a gauge.

Since regulators are probably the most important part of an oxyacetylene welding unit, they should be taken care of properly. An acetylene

QUICK»»TIP **Never use any oil on an oxygen regulator as oil is easily ignited when it comes in contact with pure oxygen. This means that you should never oil any of the regulator parts (Fig. 2.4). Almost all regulators will have bold printing on their gauges stating Use no oil.**

FIGURE 2.4

Never use any lubricant on regulator parts. Most regulators say Use no oil on the dial face.

regulator is designed to handle and control the high pressure of acetylene and make using this gas safe. Oxygen regulators perform a similar function only at much higher pressures. Both types of regulators will provide years of dependable service if they are cared for according to generally accepted procedures. Although specific instructions will be included with all new welding regulators, some general comments about care can be given here as well.

Before attaching a regulator to an oxygen or acetylene cylinder, crack the valve. *Cracking* means to open the valve slightly for a second. By doing this, you will clear the cylinder outlet of any dirt that may have accumulated since the tank was last used. By removing this dirt or dust, you will be ensuring that these particles will not get inside the regulator where they can cause the regulator to malfunction (Fig. 2.5).

After the tank valve has been cracked, the regulator can be attached. Hand tighten, being careful not to cross thread the brass fitting. After the fitting is snug, tighten it with a wrench. With the regulator turned off, slowly open the valve on the cylinder. It is important that you do not open the cylinder valve too quickly. The pressure from the tank can damage the regulator if it is allowed into the regulator all at once. After opening the valve, check for leaks around the fitting.

Probably the best way to determine that the connections are snug is to use the soap test. Simply mix up a small amount of dish soap and water. Then put a few drops of this foamy mixture on the exposed threads of the fitting. If a leak is present, the soap mixture will bubble. If the connection is tight, there will be no bubbles (Fig. 2.6).

Leaks around regulators are often the result of not tightening the fitting enough. In other cases, there may be a small particle of dirt on the regulator nipple, preventing a tight seal. Remove the regulator, and wipe the fitting with a clean, dry cloth. Always turn off the tank valve before tightening a loose or leaking connection.

When regulators will not be used for a period of several weeks, you should relieve the pressure on the pressure valve seat. After the regulator has been disconnected from the cylinder, turn the pressure-adjusting screw until it moves freely. In most cases this will mean screwing the adjusting handle out from the body of the regulator. The reason for doing this is to relieve the pressure on the valve seat and prolong the life of the regulator unit (Fig. 2.7).

The regulator valve seat should be kept clean at all times. Some regulator units come with plastic covers that are fitted over the regulator intake and outlet orifices. These do a lot to keep the regulator connecting fittings clean. In the absence of these covers, the threads of the fittings should be wiped clean with a dry cloth before making the attachment to the tank and fastening the hose connections. While wiping these fittings, it is also a good idea to inspect them for signs of wear or damage to the threads. Damage to the threads will prevent a secure connection, and the unit will leak. If this ever happens, the unit should be returned to the manufacturer or qualified welding equipment repair shop. Never use a regulator in need of repair.

There are few problems with a well-maintained regulator. It will provide years of dependable service if it is simply kept clean and not dropped or otherwise damaged. Nevertheless, there are a few problems that may develop through general use.

FIGURE 2.5

Before attaching the oxygen regulator to the tank, the valve should be opened for a moment to clear the orifice. This is known as *cracking* the valve.

FIGURE 2.6

To check for leaks, the soap bubble test is the most reliable. Simply put some of the foam on the connection, and watch for large air bubbles, which would indicate a leak. If no large bubbles appear, the connection is sufficiently tight.

FIGURE 2.7

To relieve the pressure on the regulator diaphragm, turn the control lever to the left until there is little resistance.

Regulator Freezing

The first problem is freezing of the regulator. The function of a regulator is to control high-pressure gas, which may result in the unit's becoming so cold that a layer of frost collects on the surface of the regulator. As the gas expands, it will absorb heat from the regulator and cool it. As the regulator becomes cooler, the moisture in the atmosphere condenses on the cold metal parts of the regulator and freezes.

Although frost on the outside of the regulator will not affect its operation, the gas flow may fluctuate, with a possible end result of the flame's going out. Generally speaking, if the regulator freezes, it is a direct result of moisture in the gas and not a malfunction of the regulator.

Jump

Jump is a fairly common regulator occurrence, but it is not really a problem at all. Jump usually happens when the flow of oxygen or acetylene is turned off at the torch handle. At this time, the gauge indicating the amount of pressure in the tank will rise momentarily and then stop. This short pressure rise is the jump. The amount of jump really indicates the true difference between the pressure at the regulator outlet when oxygen and acetylene are flowing and when they are not flowing. When the torch valve is closed, pressure on the regulator's gauge rises until the diaphragm is moved far enough for the regulator valve to close tightly. This closed valve regulator is called *lockup*. Jump is not a regulator malfunction. It is a normal reaction to changes in flow or no-flow pressures.

Creep

Creep is probably the only true problem with regulators. Creep can happen when the torch valve is shut off at the torch handle. The low-pressure or working-pressure gauge then moves upward. Limited creep is very similar to jump and may be hard to distinguish. In extreme cases, creep will cause the gauge to rise until either it registers the same pressure as the cylinder or a connection is blown, usually in the hose. Creep always occurs when the regulator is defective and cannot reach lockup. Needless to say, you should never operate a regulator that has a tendency to creep.

Hoses

Hoses are another important part of the oxyacetylene welding system. There are always two hoses, one for oxygen and another for acetylene. Each end of the hose has a fitting that attaches to either the regulator or the torch handle. As with all other connections, the oxygen fittings are right-handed threads, and the acetylene connections are left-handed with an identifying groove cut around the circumference of the brass fitting. All hoses are colored: green for oxygen and red for acetylene. Additionally, it is common practice, at least with new equipment, for the hose maker to print "oxygen" on the green oxygen hose. It is impossible to attach the acetylene hose to the oxygen fitting on either the regulator or torch handle because the threads, as mentioned, are different (Fig. 2.8).

SAFETY CHECKLIST

REGULATORS

___ Crack open the valve on the cylinder before installing any regulator.

___ Inspect the regulator and valve threads for damage before connecting.

___ Be certain that you are installing the proper gauge on the cylinder at hand. As a rule, acetylene connections have a groove machined on the outside while oxygen nuts are plain. Additionally, oxygen gauges commonly have green markings, and acetylene gauges usually have red markings.

___ Snug up all fittings with the proper size wrench; never use pliers as these could damage the brass nuts.

___ Open the pressure-adjusting screws all the way out before opening the cylinder valve. This will ensure that damage does not happen to the internal parts of the gauges.

___ Never use oil or grease around oxygen equipment. This includes hands, gloves, and clothing as well.

___ Open cylinder valves slowly to prevent a surge of pressure to regulator parts. It is wise to turn valves from the side rather than the top. If the valve or gauge is faulty, it will blow up from the cylinder top. Rapid opening results in tremendous pressure and heat on regulator parts and body.

___ Check for leaks with the soapy water test or by closing the tank valve and watching for a pressure drop at the gauge. If there are leaks, the gauge reading will fall.

___ Never use a regulator in need of repair. Broken glass or a tendency to creep are good reasons for bringing your regulator to a qualified repair shop.

___ Do not exceed the recommended working pressures for both oxygen and acetylene. Generally speaking, oxygen pressure should not exceed 70 psi, acetylene 15 psi.

___ Install reverse-flow check valves in the regulator outlet or torch inlet for both oxygen and acetylene connections.

Generally speaking, all hoses produced today are made from a high-quality synthetic or natural rubber material. Oxygen and acetylene hoses are flame and oil resistant and are also reinforced with a nylon mesh inside the hose. The end result is a very tough hose that is tested at the factory to withstand pressures of up to 400 psi (Fig. 2.9).

Most of the welding hoses in use today are *double,* or *Siamese,* as they are called in the trade. This simply means that the two hoses are joined along their entire length with a special adhesive. Handling a hose of this construction is much easier than it would be if two separate hoses were being used. Hoses for welding are available in several different lengths and inside diameters, the most common size being 20 feet long and 3/16 inch inside diameter.

Check your welding hose periodically for signs of wear or deterioration. It is a good idea to wipe the hoses regularly with a clean, dry cloth. You can wipe an extremely dirty hose with a damp cloth. As you are cleaning the hoses, check for signs of wear.

Older hoses can be checked for leaks by immersing them in a pail of water while they are under working pressure. The same soap and water test mentioned earlier can be used to check connections around fittings and other parts of equipment under pressure. If a leak is ever discovered, it should not be repaired with tape; instead, the hose should be taken to a qualified repair shop or replaced.

While working, it is a good idea to never step on the hose. You should also make certain that the hoses never lie where they can be

FIGURE 2.8

The hose on the left is for acetylene (notice the grooved nut) while the hose on the right is for oxygen.

FIGURE 2.9

Hoses for welding are specially designed to withstand high pressure.

QUICK›››TIP A good-quality welding hose will last for many years with a minimum of care. Before using a new hose, the talcum powder, which all new hoses are lined with, should be blown out with compressed air. Failure to do so may cause some of this powder to enter the regulator or torch handle where it can cause a malfunction. Probably the easiest way to do this is to bring the new hose to your local gas station and use the air hose to blow out the lines—unless, of course, you have access to compressed air in your workshop.

FIGURE 2.10

When storing hoses for a prolonged period, it is a good idea to cover the ends with a piece of tape. This will keep dust, dirt, and insects out of the hoses.

damaged by hot sparks or slag. It is a good practice when storing welding hoses for a long period of time to put a piece of tape (masking tape works well) over the fittings to prevent the entrance of dirt or insects. Sealing off the hose fittings will ensure that your hoses remain internally clean and clear of obstructions that could damage equipment (Fig. 2.10).

Torches

The business end of any oxyacetylene welding outfit is the *torch,* or, to use the correct terminology, the *blowpipe.* The oxygen and acetylene pass through their respective regulators, through the hoses and into separate fittings in the handle of the torch. A control knob for each regulates the flow through the torch handle and, in fact, controls the flame (Fig. 2.11).

At this present time, there are two types of torches currently available: *injector* and *medium pressure.* The major difference between the two is that the injector can use acetylene at pressures of less than 1 psi while the medium pressure torches require that acetylene be supplied in pressures from 1 to 15 psi. We will discuss both types of blowpipes (Fig. 2.12).

Injector Blowpipes

Injector blowpipes, often called *low-pressure blowpipes,* are all similar in design but will differ slightly from one manufacturer to another. Generally speaking, they all work as follows. The rear of the torch handle has two hose connections, one for oxygen and another for acetylene. The front of the handle forms a chamber where the oxygen and acetylene meet for the first time and are mixed before passing through the tip to be burned.

As oxygen passes through its tube in the handle and into the mixing chamber, it creates a suction action that draws acetylene along and into the mixing chamber as well. Here the oxygen and acetylene are allowed to expand slightly and to be thoroughly mixed. As the name "low pressure" implies, these blowpipes have the ability of working under very low acety-

FIGURE 2.11

Control knobs on the torch handle regulate the flow of oxygen and acetylene.

lene pressures and are handy for light-duty work such as wire sculpture. They are also commonly used in industry where acetylene is supplied from a generator. Low-pressure blowpipes can also be used with cylinders of oxygen and acetylene as well (Fig. 2.13).

Medium-Pressure Blowpipes

Medium-pressure blowpipes, as mentioned earlier, require acetylene with a minimum working pressure of 1 psi, up to about 15 psi. The major difference between a medium-pressure blowpipe, sometimes called a *balanced-pressure blowpipe,* and a low-pressure blowpipe lies in how the oxygen and acetylene are mixed. Although specific details of this function will differ among manufacturers, they all operate in the same basic manner. Inside the blowpipe handle there will be a large hole in the center surrounded by several smaller holes. All of these holes are located just behind the mixing chamber and base of the tip. Oxygen passes through the center hole, and acetylene exits through the surrounding smaller holes. Some medium-pressure blowpipes are exactly opposite, but the basic principle remains the same. The gases are then allowed to expand momentarily so thorough mixing can take place before passing out the tip to be burned. In all cases medium-pressure blowpipes are designed to operate at equal pressures of both oxygen and acetylene. The manufacturer of the blowpipe will always supply instructions and recommendations for both oxygen and acetylene pressures so that a specific flame can be developed at the tip (Fig. 2.14).

Medium-pressure blowpipes are probably the most common type of welding torch in use today. They provide a welding tool as versatile as possible within the limitations of oxygen and acetylene (Fig. 2.15).

Torch head   Barrel

Torch body

Control
valves

FIGURE 2.12

Diagram of a welding blowpipe.

Key
■ Oxygen
▦ Acetylene
▨ Mixed gases

FIGURE 2.13

Internal diagram of a low-pressure, or injector, blowpipe.

Key
▦ Oxygen
■ Acetylene
▨ Mixed gases

FIGURE 2.14

Internal diagram of a medium-pressure blowpipe, sometimes referred to as a balanced-pressure blowpipe.

Tips

No welding blowpipe is complete without a welding tip. There are many different sizes of tips, and each torch maker offers a line of various tips designed to be used with their particular unit. Basically all welding tips are made the same. They are most commonly made of a

FIGURE 2.15

Medium-pressure blowpipes are probably the most common type for general-purpose welding.

FIGURE 2.16

Different size welding tips can be used with the same torch handle.

heavy-gauge copper tubing that has a high resistance to reflective heat. Some tips are straight while others are curved. In principle, all welding tips are simply a piece of tubing with a threaded end that attaches the tip to the torch body. The other end has a hole through which the oxyacetylene mixture flows and is burned (Fig. 2.16).

All welding torch makers offer tip size suggestions for the various thicknesses of metals being welded with their particular blowpipe. Basically, the thickness of the metal will determine the tip size. The tip size being used will have a bearing on the pressure settings for both oxygen and acetylene (Table 2-1).

Most welding blowpipes have a cutting attachment that fits on the torch handle in place of a welding tip. A cutting blowpipe (often called the *head* of an oxyacetylene blowpipe) can be either low pressure or medium pressure depending, of course, on the basic type of unit. All cutting blowpipes are designed to allow mixed oxygen and acetylene to be released through special preheated orifices in the head of the unit. All cutting tips are the same, though the holes may be of different diameters. Basically, there are a series of small holes surrounding a larger hole in the center of the tip. The oxygen and acetylene mixture burns at the smaller holes. The center hole allows for a stream of pure oxygen to be added to the flame by the welder simply by pressing the long lever on the cutting attachment handle (Fig. 2.17). It is this flow of pure oxygen that does the actual cutting after the metal has been preheated by the cutting torch flame (Figs. 2.18 and 2.19).

Table 2-1   Pressure Chart

Metal Thickness (inches)	Tip Size No.	Size of Welding Rod (inches)	Oxygen (psi)	Acetylene (psi)
1/32	1	1/16	5	5
3/64	2	1/16	5	5
1/16	3	1/16	5	5
3/32	4	3/32	5	5
1/8	5	3/32	5	5
3/16	6	3/32	6	6
1/4	7	1/8	7	7
5/16	8	5/32	8	8

Cutting Attachments

A cutting head can be either an injector or medium-pressure type. Obviously only the proper type of cutting attachment should be used with the torch handle you have. A low-pressure cutting head is attached to a low-pressure blowpipe, for example. The operation of both types is the same as for the respective welding handles covered previously.

Just as there are different sizes of welding tips, there are also different sizes of cutting tips. Each torch maker recommends certain cutting tip sizes for various thicknesses of metal. Oxygen and acetylene pressures should be adjusted accordingly (Table 2-2).

Slag and Flashback

We have discussed both welding and cutting tips in this chapter. Thus, this is a good time to mention a few things that should be done to all tips. Because the cutting or welding tip is necessarily very

FIGURE 2.17

The oxygen cutting lever is located on top of a standard cutting torch.

FIGURE 2.18

The cutting attachment fits into the torch handle in place of standard welding tips.

FIGURE 2.19

A standard cutting attachment tip has six preheat holes surrounding a larger oxygen cutting hole in the center.

close to the work, it may become clogged with *slag*—that is, tiny bits of molten metal. When this happens, the flame will sputter and, in extreme cases, possibly go out. If the flame begins to spit and spurt, the only solution is to clean the tip with either a special tip-cleaning tool or a suitable size drill bit. It is important, when cleaning any welding or cutting tip, to move the tool in a straight in-and-out motion so the hole is not enlarged or made bell shaped (Fig. 2.20).

If a welding or cutting tip should ever become damaged, as a result of bending or cross threading, it should be replaced.

Table 2-2 Pressure Chart for Cutting Tips

Metal Thickness (inches)	Tip Size No.	Oxygen (psi)	Acetylene (psi)
3/8 – 5/8	0	30 – 40	5 – 15
5/8 – 1	1	35 – 50	5 – 15
1 – 2	2	40 – 55	5 – 15
2 – 3	3	45 – 60	5 – 15
3 – 6	4	50 – 100	5 – 15

TRADE SECRET

Often, especially when the tip of the blowpipe is placed too close to the work, the face of the tip, in addition to becoming clogged, will form a buildup of metal. This buildup should be removed with an abrasive such as a file or emery cloth. It is important to work carefully to remove carbon and slag deposits whenever they happen (Fig. 2.21).

Usually a small bend can be repaired by simply placing the tip on a block of wood and striking gently with a plastic or leather mallet. Care must be exercised when doing this, however, so that the copper tubing is not compressed (Fig. 2.22).

If a torch tip should become clogged, as a result of slag buildup, the burning oxygen and acetylene may flow back into the hose and regulator, causing what is known as a *flashback.* Special fittings have been developed to reduce the danger of flashback. If your torch does not have them, you should consider the investment. These fittings are usually attached between the end of the hoses and the torch handle. They will close off the flow of oxygen and/or acetylene whenever there is a flashback and the flow of gas is reversed. Many of the new welding outfits on the market today have these special fittings included with the unit (Fig. 2.23).

Clothing and Gear

In addition to oxygen and acetylene tanks, regulators, hoses, torch, and an assortment of tips, the do-it-yourself welder will need some safety clothing and gear including goggles (with lenses designed for gas welding), leather gloves, a long-sleeved shirt, heavy pants without cuffs, boots, and, possibly, a hat. Protective clothing is covered in detail in Chapter 14 on safety, so it will not be dealt with here. One other item the welder will need is some type of ignition system to light the torch. Matches are definitely out of the question.

Probably the most common type of torch lighter in use today is the one shown in Fig. 2.24. In theory, this lighter produces a hot spark every time the flint is rubbed across the file in the bottom of the cup. In practice, however, this type of torch lighter is not worth the price of the steel from which it is made. Unfortunately most welding outfits—regulators, hoses, torch cutting heads, and tips—usually come with one of these lighters. In my opinion, the first thing you should do when you purchase your welding outfit is to throw out the striker that comes with it (if it looks like the one in Fig. 2.24) and buy a good one instead. Welding supply houses, the

FIGURE 2.20

Welding tips will become clogged with normal use. While cleaning tools are available, the proper size twist drill bit works just as well if it is used correctly.

FIGURE 2.21

Clean the end of a welding tip by rubbing it across emery paper.

FIGURE 2.22

Straighten bent welding tips with light blows from a hammer. Use a wooden block underneath.

FIGURE 2.23

Flashback arrestors are attached to the hose end of the torch handle and eliminate the possibility of a flashback. Most new torches come with these special safety fittings.

same places where you buy your acetylene and oxygen, will often carry several different types of strikers.

One relatively new addition to the lighter line is an electronic striker. This type does not rely on the old principle of a piece of flint and steel but instead uses a tiny hand-operated generator to throw a spark between two terminals. The heart of this little unit is a tiny magneto, similar to the magneto in old-time Model T Fords and older tractors. Every time the handle is squeezed, a series of sparks jumps between the terminals. The result is that the torch can be quickly ignited (Fig. 2.25).

Wrenches

As a rule, all connections on the tanks, hoses, and blowpipe should be tightened with a suitable size wrench. All new welding outfits should come equipped with a special wrench for snugging up connections. If your welding gear is older or if the special wrench has been lost, use a substitute open-end wrench. Most experts will agree that an adjustable wrench or pliers are poor choices for tightening connections. The main reason is that these tools do not grip as well as wrenches that are made for the purpose and can, therefore, damage the fittings.

The standard striker supplied with most welding outfits.

This new type of torch lighter produces about 10 bright sparks every time the handle is squeezed.

Setting Up an Oxyacetylene Outfit

This chapter will explain how to set up the oxyacetylene welding equipment. We will also cover some of the basics of torch adjustment and standard safety practices. More specific information such as how to braze, weld, and cut metal can be found in later chapters.

Attaching Regulators and Hoses

To set up for oxyacetylene welding, we begin by making certain that the tanks of oxygen and acetylene are securely fastened to an immovable object, such as a pole or a cart specifically designed to hold the tanks. All welding supply houses carry a line of hand trucks for this purpose. They are a good investment because they safely hold the tanks and give the welder mobility as well (Fig. 3.1).

Next, attach the regulators to the tanks. First, crack the oxygen tank. Open the valve slightly so that some of the oxygen escapes and clears the valve orifice. Wipe the oxygen regulator fitting with a clean, dry cloth, and then fit the regulator into the valve on top of the tank. Tighten the regulator fitting by hand. Then use a special wrench to make sure the connection is tight. After you secure the oxygen regulator, attach the acetylene regulator to the acetylene cylinder valve in the same manner. Remember that the acetylene fitting is left-handed and must be turned in the opposite direction of the oxygen fitting. As a

FIGURE 3.1

A cylinder hand truck is probably the best way to securely store or move oxygen and acetylene tanks.

reminder, note that the acetylene fitting has a notch around the outside of the securing nut (Figs. 3.2, 3.3, and 3.4).

After you have attached the regulators to the proper cylinders, attach the hoses: the green one to the oxygen tank and the red one with the left-handed threads to the acetylene tank. If the hoses are new, you should blow them clear with compressed air first. Attach the other ends of the hose to the respective fittings on the blowpipe handle. Make certain that all connections have been secured with a wrench. At this point, the valves on both tanks should be turned to the off position, as should the regulators and the controls on the torch handle (Fig. 3.5).

Opening Valves

The next step is to very slowly open the oxygen cylinder control valve, allowing oxygen to flow into the regulator. You must do this slowly so there will not be a surge of pressure on the regulator diaphragm. Once the gauge begins to register pressure, open the oxygen valve 1 to 1½ turns. Then open the control valve on the acetylene tank. This valve turns to the left and should also be opened slowly, not more than 1 full turn (Fig. 3.6).

After all connections are tight and the oxygen and acetylene have been turned on, you should give a quick check to make certain that there are no leaks. Since you will not be able to see leaks, you must check by listening or smelling all of the connections. If you are ever uncertain as to whether or not a connection is leaking, use the soapy water test.

Setting Working Pressures

After you are sure that all connections are secure, you can set the working pressures in the hoses. This is done by turning the control levers on the face of the regulators. You will have to consult the chart that came with your welding outfit to determine the proper working pressure for the tip you are using (Fig. 3.7).

While setting the working pressures, it may be necessary to open the corresponding control knob on the torch handle. This will clear the lines and enable you to check that oxygen and acetylene are, in fact, flowing through the handle of the blowpipe. After you have set the working pressures—work on one line at a time—turn off the control knob in the handle and watch the regulator gauges. They should move only slightly if at all. You must set the working pressure for both oxygen and acetylene according to the tip chart for your particular welding equipment.

Lighting the Torch

After the proper working pressures have been set, you can light the torch. Before you do, however, you should put on a pair of gloves and goggles. You can raise the goggles to your forehead, so that they can be easily pulled down over your eyes when you need them. You should also be standing in front of your workbench, with the hoses coming from behind you so that there is no possibility of their

FIGURE 3.2

Crack the oxygen valve.

QUICK»»TIP

When setting the pressure in the hoses, with the gas or oxygen flowing into the atmosphere, you must make certain that there is no open flame or other heat source in the area that could cause the fuel to ignite.

being hit and damaged by a hot spark or slag. Assuming that you are right-handed, hold the blowpipe in this hand and the striker in your left. Turn the acetylene control knob open a crack, just enough to let some gas flow through the tip. Hold the striker close to the tip and squeeze it a few times to generate enough sparks to light the acetylene (Fig. 3.8).

FIGURE 3.3

The regulator fittings should be wiped with a clean dry cloth before attaching to the tank valve.

FIGURE 3.4

After you have snugged up the regulator fitting by hand, tighten it securely with a wrench.

Acetylene Flame

The first flame that appears will be yellow to orange in color, and black smoke will be present at the end of it. This flame is called the *acetylene flame,* and it does not contain any pure oxygen. If too much acetylene is present, the flame will be a few inches away from the hole in the tip, possibly roaring like a jet engine. In extreme cases, this flame will blow itself out. To correct, simply reduce the amount of acetylene flowing through the blowpipe handle. When the acetylene flame is about 8 to 10 inches long, with only a small amount of black smoke, begin to introduce oxygen by slowly opening the oxygen control knob on the torch handle (Fig. 3.9).

Carburizing Flame

As soon as pure oxygen is added to the acetylene flame, some changes begin to happen. First the black smoke will totally disappear and the flame will change from a yellow-orange to a whitish color. There will also be three distinct parts to the flame as seen in Fig. 3.10. The flame produced when oxygen is first introduced is called the *carburizing flame.* This flame consists of a greater amount of acetylene in relation to oxygen and therefore is not hot enough for welding. It can be used, though, to add carbon to a weld or to blacken the surface of metal.

Neutral Flame

As more oxygen is added to the flame simply by turning the oxygen control knob, it will change again. The excess acetylene feather will appear to shrink toward the tip and will finally disappear into the small white cone at the base of the tip. This flame is called the *neutral flame,* and it is used to weld steel and braze most metals. There are only two parts to this flame, a sharp bright cone at the end of the welding tip and a heat envelope (Fig. 3.11). The cone should be well defined and white in color.

Oxidizing Flame

If the oxygen control knob is opened more, the nature of the neutral flame will change once again. The new flame is called an *oxidizing flame.* This flame is often used for brazing. To adjust a neutral flame to an oxidizing flame, you must increase the ratio of oxygen to acetylene. You can either reduce the amount of acetylene coming out of the blowpipe or increase the amount of oxygen (Fig. 3.12).

FIGURE 3.5

Tighten hose connections to the regulator with a special wrench.

An oxidizing flame should never be used for welding. The excess oxygen will cause breakdowns and defects in both the metal being welded and the weld itself.

When you first begin welding with an oxyacetylene outfit, you should practice adjusting the blowpipe to the different types of flames: acetylene, carburizing, neutral, and oxidizing. Once you have become proficient at adjusting to these flames, especially the last two, you can then work on welding technique.

Concluding Tasks

After you have finished welding, you will want to turn off your blow-pipe. Always turn off the acetylene flow first. Oxygen will continue to flow through the torch, ensuring that the orifice is blown clean. After a few moments, turn off the oxygen.

FIGURE 3.6

It is important to open the valve slowly after the regulator has been securely fastened to the tank. This will ensure that there is not a surge of oxygen to the internal parts.

TRADE SECRET The name "neutral flame" is derived from the fact that the chemical effect of this flame on molten metal during the welding process is neutral. This is true, of course, providing that the flame is held in the proper manner with the inner cone not quite touching the molten metal. It will be to your advantage to learn how to adjust your blowpipe to arrive at a neutral flame, as it will be the most frequently used flame in all of your welding work. You can adjust the length of the cone by adding more acetylene and then more oxygen. The neutral flame is important not only because it is the most widely used flame, but also serves as a basis for adjusting to or from other flames used in welding and general metalworking.

If you are stopping work for just a short period of time, say, 10 minutes or more, you should close the tank valves, leaving oxygen and acetylene in their respective hoses. If you are stopping work for a longer period, you should, in addition to closing the tank valves, bleed the line.

To bleed the hoses in your unit, begin by turning off the flow of oxygen and acetylene into the regulators. Then open the acetylene control knob on the blowpipe handle and let the acetylene in the hose escape. Keep an eye on the gauge of the regulator. As the pressure in the hose drops, the needle will swing to zero. Once this happens, close the control knob on the handle and repeat the operation, bleeding the oxygen line.

Another task you must accomplish when shutting down your welding outfit, after the hoses have been bled, is to release the regulator pressure handles so that there is no tension on the diaphragms. Do this simply by turning the handle on the face of the regulator until there is very little or no resistance. This small act will do a lot to prolong the life of your regulators.

FIGURE 3.7

Set the working pressure in the hose by turning the regulator control lever to the right. The oxygen working pressure is being set at about 30 psi while the pressure in the tank is about 1,700 psi.

FIGURE 3.8

Hold the striker close to the torch tip, with the acetylene flowing. Light the gas by striking a spark into the gas.

Bright, bushy yellow flame

Tip

FIGURE 3.9

A pure acetylene flame containing no oxygen.

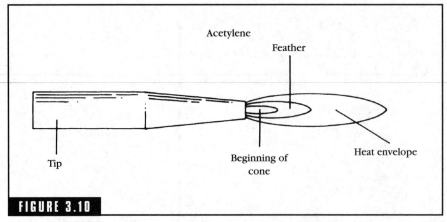

FIGURE 3.10

As soon as oxygen is introduced to the acetylene flame, a cone begins to form and the flame takes on a definite shape. It is called a *carburizing flame*.

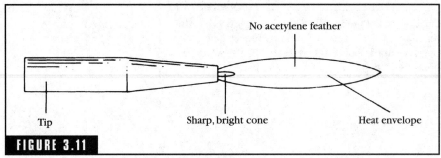

FIGURE 3.11

As more oxygen is added, a neutral flame results. This flame is the most commonly used for all welding.

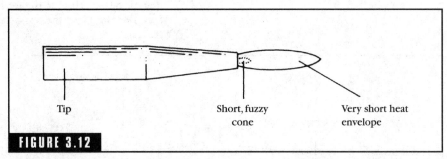

FIGURE 3.12

If more oxygen is added to the neutral flame, the cone will decrease in size. The result is an oxidizing flame.

Other Types of Welding and Equipment

U p to this point, we have covered only the equipment necessary for oxyacetylene welding and cutting. The bulk of this book will, in fact, concern itself with this method of joining metals. I believe it is the easiest type of welding for the beginner to learn, and it is versatile enough to enable the do-it-yourselfer to perform most types of metal joining tasks. Remember that all welding falls into three rather broad categories: gas welding, electric welding, and gas-electric welding. Since we have already covered the equipment necessary for the first type, it is now time to turn our attention to the others.

Arc Welding

If you feel that you would like to try your hand at *arc welding*, I suggest that you contact your local welding supply house for more information and possibly a demonstration. There are also courses offered at vocational schools and adult education classes. In truth, arc welding is not difficult, but it does require a certain discipline and strict adherence to specific safety practices. Learning the basics of arc welding is best accomplished under the tutelage of a knowledgeable expert.

Shielded metal arc welding, commonly referred to as simply *arc welding*, is probably the most common form of welding in use today. It is extremely attractive for commercial work because it is fast and produces strong welds. The major application of arc welding is for joining mild-carbon and low-alloy steels.

Electrodes

The underlying principle of arc welding is quite simple. An electric welding machine produces a specific amount of current, which passes through a cable, called an *electrode lead,* to a special handle that holds an electrode. An electrode is similar to a welding rod used for gas welding except that it must also conduct electricity. Electrodes are also covered with *flux.* An arc is struck by moving the electrode close to the metal being welded. The arc is extremely hot, enough to melt both the end of the electrode and the edges of the metal. The current passes through the electrode, melting the end of the metal where the arc from the electrode touches the metal. A *work-lead cable* connects at the welding machine and is also clamped to the metal being welded, making it possible for the electric current to complete its circular path (Fig. 4.1).

The arc creates a very intense heat that melts the edges of the metal and the end of the electrode at the same time. The melted metal from the electrode moves across the arc and is deposited on the metal, forming a strong weld, in most cases equal to or stronger than the base metal itself.

The flux coating on the electrode produces an inert gas when an arc is struck, and the gas shields the arc and the weld from the surround-

FIGURE 4.1

An oversimplified diagram of a typical arc welding unit.

FIGURE 4.2

As the electrode covering melts, a special shielding gas is produced. This gas protects the molten metal from the atmosphere.

ing atmosphere. At the same time deoxidizers are produced that purify the electrode metal. This flux also forms a slag, which protects the molten metal from oxidation. After the weld has cooled a bit, the slag is commonly removed with a special tool (Fig. 4.2).

As you can well imagine, electrodes are a very important part of an arc welding system. They are commonly identifiable by a special numbering system developed by the American Welding Society (AWS) according to strength, welding position, usability, and analysis of deposited metal (Fig 4.3).

Probably hundreds of different flux-coated electrodes have been developed to perform under various applications. Electrodes are available in sizes ranging from $1/16$ to $5/16$ inch diameter and from 9 to 18 inches long. Electrodes 14 inches long are probably the most widely used in the welding industry.

Current-Supplying Machine

Arc welding is made possible by a machine that supplies either *direct current* (dc) or *alternating current* (ac). Needless to say, the current

FIGURE 4.3

A selection of electrodes.

must be controllable and adjustable according to needs. At one time all
arc welding was done with dc because ac was unsuitable. Now, however, many welding equipment companies have come out with a line of
ac welding units that will weld a wide range of thicknesses and types
of metals (Fig. 4.4). One advantage of ac arc welding units is that they
can be produced at a lower cost, which makes them more affordable for
arc welders. The ac transformer is probably the smallest, lightest, and
least expensive of all. It is also relatively quiet during operation.

An arc welding machine, either ac or dc, provides the electric power
in the proper current voltage to maintain the welding arc. The higher
the amperage, the greater the electric flow that will be transferred
through the electrode to the metal being welded. Also, the higher the
current, the faster the electrode will be consumed. Generally speaking,
the higher settings on the welding machine are reserved for welding
thicker base metals.

Protection Gear

Arc welding requires the welder to wear a special helmet to protect the eyes from the extremely bright arc and to protect the face and head from flying molten metal particles. It is also standard practice to wear heavy gauntlet gloves and leather clothing. The clothing and protection gear are often bulky and confining, but they are necessary to provide a line of defense against burns and eye damage (Fig. 4.5).

A relatively new helmet for arc welding has been developed by several manufacturers. The auto darkening arc welding helmet has a lens that is slightly tinted. When an arc is struck, however, the lens darkens instantly to protect the user's eyes. Once the arc is no longer bright, the auto darkening lens returns to the slightly tinted shade, and the user can see normally.

FIGURE 4.4

A typical arc welding outfit.

You can adjust all auto darkening lenses for tint, usually from a 9 to 12 tint. You can also adjust the amount of time that the lens stays darkened after the bright arc ceases. Auto darkening arc welding hoods are expensive—prices start at about $250—but they are a good investment if you plan to do much arc welding.

Striking an Arc

While it is not my intention to cover arc welding in detail, I would like to mention some of the basics of this form of metal joining. The first step is to learn to strike a smooth arc without allowing the electrode to actually touch the metal. When this happens, the electrode will stick to the metal, which is known as *freezing.* If this ever happens, there are two courses of action, one of which should be initiated immediately.

The first thing you can try when an electrode has frozen is to give a firm twist to the electrode holder. If a fast twist with an upward pull does not free the electrode, release the electrode from the holder and break the circuit. After the metal and electrode have cooled, the electrode can easily be removed from the metal with a pair of pliers, using

FIGURE 4.5

Standard arc welding face mask or shield.

QUICK»TIP Because of the possibility of freezing and the potential damage to equipment, it is a sound idea to practice striking an arc a few times with the power turned off. Once you have the general idea, then you can proceed with the power on.

---

a twisting motion. Failing this, try striking the base of the electrode, where it is joined to the metal, with a chipping hammer.

To strike a proper arc, begin by setting the recommended current on the arc welding machine. Then fasten an electrode in the holder. Hold the electrode almost straight above the metal to be welded and tilted at about a 10-degree angle. If you are using an ac machine, you strike an arc using the scratching method (Fig. 4.6). This is similar to striking a kitchen match. If you are using a dc machine, the accepted method of striking an arc is to tap the electrode to metal rather than scratch the surface of the metal (Fig. 4.7).

In any case, as the electrode is brought close to the work, an arc will begin to form. As soon as you see a flash, you should raise the tip of the electrode up from the work until it is about $\frac{1}{4}$ inch above the surface and the arc is still present. If the arc is held at this height in a steady position, a puddle of molten metal will begin to form where the arc touches the parent metal. At this time the electrode should be lowered toward the metal until it is at a height roughly equal to the thickness of the electrode. For example, if you are using a $\frac{1}{8}$-inch-diameter electrode, the tip should be about $\frac{1}{8}$ inch above the work. This, incidentally, is about the time that beginners run into a freezing problem as mentioned earlier.

Striking a proper arc should be practiced until an arc can be struck, raised, and then lowered into welding position without creating a short circuit and freezing the electrode. One very real indication that a proper arc has been struck is the arc will give off a crackling sound, very similar to bacon frying. This assumes, of course, that the welding machine has been set properly for current based on the electrode being used.

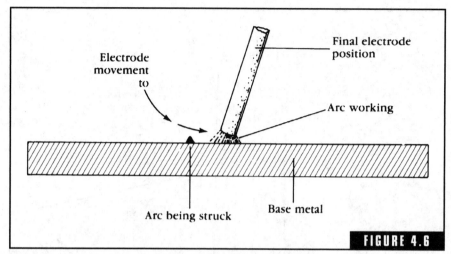

Electrode movement to

Final electrode position

Arc working

Arc being struck

Base metal

FIGURE 4.6

Use the scratch method to strike an arc with an ac arc welder.

Cutting Metal

Arc welding equipment can also be used for cutting metal. As with arc welding, arc cutting uses a powerful electric force to melt the metal. The molten metal is then either oxidized or blown away with a special attachment that introduces compressed air. Generally speaking, special electrodes are used for arc cutting. These electrodes are coated with a special insulating material that does not conduct electricity. This coating allows the welder to touch the metal being cut without the possibility of freezing the electrode. The coating on a cutting electrode acts as an arc stabilizer, which concentrates the arc and at the same time intensifies the arc's action. Cutting electrodes are nonconsumable, which means that they will retain their original diameters and lengths during cutting.

Arc welding and cutting, as mentioned earlier, are not difficult to learn. They should not be attempted, though, without knowledgeable supervision.

Tungsten Inert Gas Welding

Tungsten inert gas welding, or simply *TIG welding,* as it is called in the trade, was first introduced in the late 1940s and was an answer to welders' prayers. One of the problems when any two metals are joined

Electrode moving down

Final electrode position

Arc struck

Arc working

Base metal

FIGURE 4.7

Use the tapping method to strike an arc when using a dc arc welder.

with heat is that the molten metal is vulnerable to contamination from oxides in the atmosphere. The end result is a weld that may have weak spots. Before TIG welding was introduced, welders had to resort to various fluxes to remove contaminated material from the weld.

With the introduction of TIG welding, the problem of potentially weak joints and welds was eliminated. Basically, TIG welding is arc welding with the addition of a special inert gas, in most cases argon and/or helium, which shields the area being welded from airborne contaminated material. The gas itself is inert, which means that it has a neutral effect on molten metal. The end result is a very clean, strong weld.

Tungsten inert gas welding differs in another way from standard arc welding in that the electrode used to strike an arc is nonconsumable;

it does not decrease in size or length. The heat from this tungsten electrode is sufficient to melt the edges of the metal being welded, but it is not hot enough to melt the tungsten. A welding rod, not unlike an oxyacetylene welding rod, is used as a filler material and is not attached to the electrode in any way.

Tungsten inert gas welding was originally developed specifically for welding manganese, aluminum, and stainless steel. It is widely used to this day for nonferrous metals. In the process of developing a new means of welding these metals—which are used extensively in the aircraft industry—the inventors also developed a method that was fast, easy to learn, and more economical than oxyacetylene welding. Tungsten inert gas welding also produces clean, beautifully welded joints.

Unfortunately, the equipment necessary for TIG welding can be above the financial means of the casual, do-it-yourself welder. However, it is widely used in industry and specialty welding shops.

Gas Metal Arc Welding

To say that the introduction of TIG welding opened the door to new welding methods is an understatement at the very least. Unfortunately, TIG welding has its limitations, the major one being that it is not suitable for welding metals with a thickness greater than about $\frac{1}{4}$ inch. *Gas metal arc welding* (MIG welding) was originally developed as a solution to welding metals thicker than $\frac{1}{4}$ inch using the TIG principle of shielding the area being welded with an inert gas. Gas metal arc welding, when first introduced, was limited to rather thick welding applications. In a short period of time, certain refinements were introduced to the basic MIG system, which transformed this type of welding into a very versatile means of joining metals.

Gas metal arc welding differs from TIG welding in one major way: The electrode is consumed during the welding process. Actually, the electrode in a MIG system is a continuous bare metal filler wire that is constantly fed through the handle of the unit. The wire makes continuous welding possible. This filler wire is fed into the arc and melts on its way to being deposited in the weld. As with TIG welding, MIG welding produces a strong, clean weld. Carbon dioxide is used in MIG welding as the shielding gas. Figure 4.8 shows a MIG welding unit in operation.

FIGURE 4.8

Gas metal arc (MIG) welding equipment in use.

Since the introduction of TIG welding in the 1940s and the later development of MIG welding, there have been newer additions to the welding field. American know-how knows no bounds when it comes to developing a new means for accomplishing a specific task. There are probably another dozen welding processes in use today in addition to the ones covered in this chapter. But these are all used for specific tasks for welding in industry and, as with both TIG and MIG, are not really designed for the home welder in either price or application categories.

One possible exception to this is welding equipment that uses a continuous wire for welding. *Wire-feed welding* is commonly performed with a hand-held gun that contains the end of a length of wire. Electric current passes though this wire onto the area to be welded. This type of welding is considered arc welding, and it can be either shielded or unshielded, depending on the type of wire used.

A number of welding equipment manufacturers now offer wire-feed arc welding units for do-it-yourself and light commercial welding applications. Many of these units are designed to be used with standard 115-volt household current. All units are equipped with infinite wire speed control to accurately select the proper wire-feed rate needed for various welding conditions.

One of the real advantages of wire-feed arc welding units is that they are easy to use for the beginner and relatively inexpensive—under $300. Some units use flux-coated wire, and more advanced models use argon gas (for shielding) and bare wire.

In all probability the do-it-yourself welder will learn how to weld using a basic oxyacetylene welding outfit. This makes a lot of sense on several levels. The oxyacetylene process has the ability to handle a wide range of welding projects, certainly all of those encountered by the casual welder. Oxyacetylene welding and cutting is also a bit safer than the other types such as arc welding. There is less chance of injury to the user. Another strong point is that the equipment necessary to weld with gas is affordable. There are many welding outfits on the market that can be purchased for under $200 (Fig. 4.9).

FIGURE 4.9

The typical welding outfit will contain many of the tools and accessories you will need for repairs and building projects around the home and farm.

Metals and Their Properties

Before you can expect to achieve success in welding, you must have an understanding of the different properties of various metals. Since they differ widely in strength, composition, and melting points, it is impossible to join any two pieces of metal simply by clamping them together and heating them up to the melting point until they join. Perhaps the following very brief history of metals will help you gain a more thorough understanding of metals in general.

History of Metals

Early human beings, by chance, discovered that certain earths, when exposed to extreme heat, left a residue after cooling that was very different from stone or the original earth. They later discovered that this hard material could be pounded into various shapes that were useful for gathering food and making weapons.

The actual chemistry of metal was not really understood until fairly recently. Basically, metallic ores are oxides—that is, compounds that are composed of metal and oxygen atoms. When these ores are heated with a fuel that gives off carbon monoxide—coke or charcoal, for example— the ore loses its oxygen as a result of the heat. The end product will be metal and slag. Scientists call this process *reduction*. When the ore atoms are rearranged and the oxygen removed, the end residue is metal.

Copper

Copper was probably the first metal to be discovered. It is very common and can be heated out of ore with a relatively low temperature. Unfortunately pure copper is too soft to be used for weapons or tools. If *tin* is added to the molten copper, the cooled metal will be much harder than pure copper. The amount of tin added would, in fact, give the metal certain qualities. The addition of tin to pure copper produces an alloy called *bronze.* Human beings have known about bronze since about 2000 B.C. This metal was first used for weapons, shields, bells, and utensils.

If *zinc* is added to copper instead of tin, the resulting metal will be *brass.* Today there are a number of different brass compositions. Some of these contain iron and tin in addition to zinc. As human beings learned more about metal, they discovered that bronze and brass did not hold an edge for very long. Experimentation eventually led to tools and weapons being made from a stronger metal known as *iron.*

Iron

Iron was first discovered as a residue left over from smelting other metals. Small bits and pieces of iron would be a common byproduct in the slag of copper. Eventually some enterprising individual discovered that these pieces could be heated and then hammered into various shapes. Unfortunately, this early iron turned out to be not as strong as brass or bronze.

It is estimated by anthropologists that around 1400 B.C. the Greeks discovered that heating and reheating iron, and then hammering it, would transform the iron into something far superior to bronze in its strength and hardness. It was also discovered that repeated heating of iron would cause the metal to absorb carbon from the heating fuel. This excess carbon would change the iron into a carbon iron alloy, steel, with a skin that was very hard. Today we do much the same thing to manufacture *case-hardened steel,* which is steel with a hard coating and a softer core.

As more experiments were conducted with iron, it was discovered that if red hot metal was dropped into cold water, the result would be a metal that was harder than any other. Unfortunately, this also produced a metal that was very brittle when subjected to shock. The Romans, conquerors of the world, experimented a bit further and learned that much of the brittleness could be removed from this new metal if it was reheated and then quenched at a higher temperature. The metal retained its hardness but was no longer brittle, making it ideal for weapons. Archaeologists have discovered weapons with these properties dating back to about 1300 B.C.

Steel Making Processes

It was only 300 years ago (about A.D. 1720) that people understood the role of carbon in the steel making process. A French chemist discovered that if certain amounts of carbon were added to molten iron, the resulting metal would be steel. The problem, at that time, was how much carbon should be added and how the carbon could be combined. Sir Henry Bessemer solved this problem when he invented a method of making steel which is still in use today. This discovery, in 1856, was called the *Bessemer process.* It uses a blast furnace to produce a low-grade steel that is still in great demand for various structural steel shapes.

The Bessemer process utilizes a controlled furnace that blows air through the molten iron. As this air escapes, it takes burnt gases of carbon, manganese, and silicon with it. The end result is a steel that is free of these elements. Then controlled amounts of carbon and/or other alloying elements are added to make a specific type of steel. *Alloy steels,* which we will cover later in this chapter, contain small amounts of tungsten, chromium, nickle, manganese, and vanadium. Today steel and alloy steel are made in one of three processes: Bessemer, open-hearth, and the electric arc furnace (Fig. 5.1).

Physical Properties of Metals

Most of the physical properties of a metal can be easily identified by sight. These properties include its *color* and *magnetism.*

Color

Color as a physical property describes the surface appearance of the metal. Iron looks different from aluminum, gold is different from silver, copper looks different from zinc, and so on. But there are

> **TRADE SECRET**
>
> The do-it-yourself welder should be able to identify metals as closely as possible so that he or she will be able to join them. Generally speaking, three basic areas of information, characteristics, if you will, that the welder should be familiar with are the metal's *mechanical, chemical,* and *physical* properties. These describe the metal and make it easier to identify.

Slag

Molten
steel

Air

Wind chamber

FIGURE 5.1

Bessemer converter.

also similarities. The visual test serves only to eliminate certain metals from consideration. Experienced welders and other people who have worked with metal for a long time can usually come pretty close to guessing what type of metal they are faced with. But the rest of us will have to rely on other methods for identifying metals. Nevertheless, the color test is valuable for grouping metals into possible categories.

Magnetism

All *ferromagnetic metals*—ferrous metals contain iron—have magnetic properties. In

FIGURE 5.2

All ferrous metals are magnetic.

other words, a magnet will be attracted to these metals. Actually the magnetic test serves to group all metals into one of two rather broad categories, *ferrous* and *nonferrous.* All ferrous metals are derived from iron and its alloys. Nonferrous metals are non-magnetic and contain either no iron or insignificant amounts. Some examples of nonferrous metals are copper, bronze, brass, zinc, and aluminum (Fig. 5.2).

Melting Point

The *melting point* of a metal serves as probably a more accurate identification of specific metals and their alloys. Table 5-1 lists the various melting points of metals and alloys. One very good and accurate way to determine the melting point of a metal, aside from using a special thermometer, is to use a crayon-like material that has a known melting point. These melt point indicators are sold at welding supply houses and are available in stick form (very much like a crayon), tablets, and liquids. When these materials are heated, they will melt at a specific temperature—2,800 degrees, for example. When these special melt point indicators are used, the type of metal being considered can be fairly accurately identified (Fig. 5.3).

Chemical Properties of Metals

The chemical properties of metals include whether they are subject to *corrosion, oxidation,* and *reduction.* Corrosion is the chemical attack

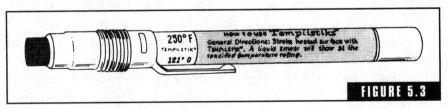

FIGURE 5.3

A temperature-indicating crayon.

Table 5-1 Melting points of metals and alloys.

Carbon	3500	
	3400	
	3300	
	3200	
Chromium	3100	
Pure iron	3000	Wrought iron
Mild steel	2900	Stainless steel, 12%
	2800	chromium
	2700	Cobalt
Nickel	2600	Silicon
Stainless steel,	2500	
19% chromium	2400	
Manganese	2300	
	2200	Cast iron
	2100	
	2000	Copper
	1900	
Silver	1800	Brass
	1700	
	1600	Bronze
	1500	
	1400	
Aluminum	1300	Magnesium
	1200	
	1100	
	1000	Aluminum alloys
	900	Magnesium alloys
Zinc	800	
	700	
Lead	600	
	500	
Tin	400	

on metal by various elements in the atmosphere and can happen only on aluminum and its alloys. Oxidation in one form is rust, and it occurs on all ferrous metals when certain oxides in or on the metal are combined with oxygen. Some metals, unless protected with a coating of some type, will combine with oxygen in the atmosphere and produce rust. The ease with which some metals combine with oxygen can complicate the welding process since pure oxygen is used. In certain instances, when oxygen combines with the oxides in metals, reduction takes place, and the volume of the metal will be reduced. Fluxes and shielding gases (such as those used in TIG and MIG welding) are commonly used to prevent reduction.

The chemical properties of metals can also be determined through certain tests. For example, chromemolybdenum, a popular aircraft steel alloy, can be identified by immersing some of its filings in a dilute sulfuric acid solution. As the filings dissolve, the solution turns dark green.

Chemical tests are not used with any frequency at the do-it-yourself level, but they are used extensively in industry to identify the composition of metals and their alloys.

Mechanical Properties of Metals

Undoubtedly the most important characteristics of metals for the welder are their mechanical properties. This is a rather broad category of descriptive terms used to identify the properties of metals. Included are *tensile strength, ductility, hardness, compressive strength, toughness, malleability, impact strength, fatigue, brittleness, stress, elasticity,* and *creep.* We will discuss each of these properties because of their importance to the modern do-it-yourself welder.

Tensile Strength, Ductility, and Hardness

The *tensile strength* of a metal is its ability to resist being pulled apart. Imagine pulling apart a piece of taffy, and you will have a good idea of tensile strength. The tensile strength of a metal is determined by placing a sample of a metal in a special machine, which is able to evenly pull with tremendous strength, and noting the point at which the specimen pulls apart. The tensile strength of a metal is expressed or measured in thousands of pounds per square inch (Fig. 5.4).

Ductility is a quality of metal that enables it to be formed or worked with force without breaking. When making wire, the ductility of the metal is important. Wire is made by pulling a metal rod through a special die that shapes the metal without breaking it (Fig. 5.5).

The term *hardness* implies solidity and firmness in a metal. Actually, hardness of a metal is its ability to resist forcible penetration of another material or metal. There are a number of methods used to determine the hardness of a metal. Probably the only test that can be done by the home welder is

FIGURE 5.4

Tensile strength of a metal indicates the amount of force necessary to pull that metal apart.

Direction of pull

Die

Die

FIGURE 5.5

A metal with good ductility can be worked with force without damage. Wire is made from metal with this quality.

a simple file test. A standard machinist's hand file is used to file the metal. The results of the filing can then be compared with the metals listed in Table 5-2.

It should be noted, when performing the file test, that several areas of the metal should be tested and an average of the findings determined. Surface defects will, of course, have an effect on the results. Therefore, extremely high or low readings should be disregarded.

Compressive Strength, Toughness, and Malleability

The *compressive strength* of a metal is another mechanical property. A metal is said to have much compressive strength if it is able to withstand a force over a large area without deforming. It is important that the force be applied slowly and evenly rather than all at once. Striking with a hammer, for example, is not a good test for the compressive strength of a metal. One example of a metal that has good compressive strength is a gasoline engine head or block. This metal can withstand high compression during operation, but it will crack or break if dropped a few feet onto a hard surface.

A metal is said to be *tough* if it can withstand repeated applications and releases of force without change. An automobile leaf spring is one example of a tough metal. A good one will stand up under the force of a heavy load and return to its original position without change for many years (Fig. 5.6).

Table 5-2 Metal Hardness

Hardness (Brinell Method)	When Filed
100	Metal is very soft, easy to file.
200	Metal reasonably soft, slight pressure required on file.
300	Metal exhibits resistance but can be filed with pressure.
400	Metal is difficult to file; hard.
500	Metal appears almost as hard as file; very difficult to file.
600	Metal cannot be filed; damage to file teeth.

The *malleability* of a metal is its ability to be worked cold without noticeable resistance. For example, low-carbon steel can be hammered or bent into shape but cast iron cannot. In this case the steel is malleable while the cast iron is not. Other examples are aluminum, copper, silver, and gold, which can all be cold rolled into thin sheets without deforming (Fig. 5.7).

FIGURE 5.6

A metal is tough if it can withstand repeated force without changing. An automobile leaf spring must be made from tough metal.

Impact Strength, Fatigue, and Brittleness

Impact strength is another metal property that is desirable in certain applications. A metal with good impact strength must be able to withstand repeated shock without fracturing or breaking down. An example of metals with high impact strength are those metals that are used for tools such as hammers and cold chisels (Fig. 5.8).

Metals are said to suffer *fatigue* when they do not hold up under continuous impact or load. Fatigue usually makes itself known in metal by a crack that lengthens in time. Metal fatigue usually happens when there are flaws in the metal combined with repeated applications of a heavy load. The flaw in the metal gives way due to the excess stress being applied (Fig. 5.9).

The *brittleness* of a metal indicates a lack

FIGURE 5.7

Malleability is the ability of a metal to be worked easily without being heated.

of ductility. The metal will break, not bend. An example of brittleness can be demonstrated with a pencil. It will not bend but will break when a certain amount of force is applied. A metal considered brittle will shatter like a piece of glass when subjected to a sudden shock (Fig. 5.10).

Stress, Elasticity, and Creep

Stress refers to the amount of force, such as a load, applied against a metal surface. Some metals will bend when stress is applied. Heat will

FIGURE 5.8

A metal is considered to have high impact strength if it can withstand repeated sharp force. Hammers and chisels are made from metals with high impact strength.

Load

FIGURE 5.9

A metal has fatigue when it cannot withstand stress from a load.

decrease the amount of stress necessary to deform metals. *Strain* is the result of stress when applied to metal.

Elasticity refers to the ability of a metal to return to its original shape after having been subjected to stress. All metals have a certain amount of elasticity, and they also have an elastic limit. The elastic limit of a metal is that point where it will not return to its original shape (Fig. 5.11).

After the elastic limit of a metal has been reached, any additional force applied will cause a permanent change in the metal's form. This is known as *creep*.

Metal Identification

As you will see in later chapters, welding methods will differ according to the type of metal being repaired or welded. It is therefore important for the do-it-yourselfer to be able to identify metals and alloys of metals. This is particularly important when working with the various types of cast iron and different carbon content steels. As mentioned earlier, the color or general appearance of a metal can be an indication of the type of metal or, at the very least, the type of metal that it is not (see Table 5-3). It will be to the new welder's distinct advantage to obtain samples of as many different types of metals as possible so he or she will have some basis for metal comparison.

Chip Test

Another way of identifying different metals is to remove a piece of the metal with a cold chisel and hammer (Fig. 5.12) and then to carefully examine the piece. This is called the *chip test*. While there are obvious limitations, it is still a valuable test that offers fairly accurate results. See Table 5-4.

Spark Test

Still another valuable test to determine the type of metal in hand involves pushing a sample of metal against a bench grinder or grinding wheel and watching the resulting sparks. As you might expect, this is called the *spark test*. It will give you a fairly accurate indication of what type of metal you have at hand. The key to the spark test is to watch the sparks as various metals give off different spark patterns. With a little experience almost anyone with a keen eye can identify different types of metals and their alloys. It should be noted however that the spark test can be used only for ferrous metals—those metals that contain iron in some form—as nonferrous metals do not have any appreciable spark.

Since the spark test is neither difficult or beyond the capabilities of the average home welder, it will be helpful to explain how it is performed. Basically, a piece of metal is lightly touched against the spinning surface of a grinding wheel. The stream of sparks that will result should then be observed against a dark background and at an angle that permits a good view. Needless to say, some type of clear eye protection should be worn while

FIGURE 5.10

A brittle metal will shatter when a heavy load is rapidly applied to it.

FIGURE 5.11

Metal that is elastic can be stretched to a certain limit while it is cold.

doing this test. The length of the stream of sparks will be directly related to how much pressure is exerted against the grinding wheel. The three things that you should be watchful for are the color, volume, and nature of the sparks (Fig. 5.13). The stream of sparks can then be

Table 5-3 Metal identification by appearance.

	Alloy Steel	Copper	Brass and Bronze	Aluminum and Alloys	Monel Metal	Nickel	Lead
Fracture	Medium gray	Red color	Red to yellow	White	Light gray	Almost white	White; crystalline
Unfinished surface	Dark gray; relatively rough; rolling or forging lines may be noticeable	Various degrees of reddish brown to green due to oxides; smooth	Various shades of green, brown, or yellow due to oxides; smooth	Evidences of mold or rolls; very light gray	Smooth; dark gray	Smooth; dark gray	Smooth; velvety; white to gray
Newly machined	Very smooth; bright gray	Bright copper red color dulls with time.	Red through to whitish yellow; very smooth	Smooth; very white.	Very smooth; light gray	Very smooth; white	Very smooth; white

	White Cast Iron	Gray Cast Iron	Malleable Iron	Wrought Iron	Low-Carbon Steel and Cast Steel	High-Carbon Steel
Fracture	Very fine silvery white silky crystalline formation;	Dark gray	Dark gray	Bright gray	Bright gray	Very light gray
Unfinished surface	Evidence of sand mold; dull gray	Evidence of sand mold; very dull gray	Evidence of sand mold; dull gray	Light gray; smooth	Dark gray; forging marks may be noticeable; cast-evidences of mold	Dark gray; rolling or forging lines may be noticeable
Newly machined	Rarely machined	Fairly smooth; light gray	Smooth surface; light gray	Very smooth surface; light gray	Very smooth; bright gray	Very smooth; bright gray

compared with Fig. 5.14 to determine the type of metal being ground.

Wrought iron is iron with no carbon present. It has a spark stream that is composed of small particles that flow away from the wheel in a straight line. The sparks become wider and brighter as they get further away from the wheel until finally they go out.

Mild steel contains a small percentage of carbon. It will have a stream of sparks similar to those of wrought iron, except that there will be, in addition, tiny forks in the sparks. The carbon causes the sparks to appear white in color.

FIGURE 5.12

The chip test is one reasonably accurate means of identifying different metals.

Tool steel will produce sparks that are similar to the mild steel sparks but with a greater number of forks in the stream. Generally speaking, the greater the amount of carbon in the steel, the more forked sparks in the trail. High-carbon steel, for example, will have a blaze of forked sparks that will begin very close to the wheel. Mild-carbon steel will have sparks with few forks that will not appear until the stream is about 1 foot away from the grinding wheel.

Cast iron will have a stream of small repeating sparks that begin very close to the grinding wheel and do not last. The spark stream will be very short.

Now that we have discussed the properties of different metals as well as how to determine what type of metal is at hand, it will be helpful to talk about the actual metals and their alloys. The metals that the average home welder will come in contact with are iron, steel and its alloys, copper, aluminum and its alloys, and magnesium. A brief description of these metals follows.

Iron

As explained earlier in this chapter, iron has been in use for quite a long time. Iron is extracted from iron ore and is the second most common metal on earth. Aluminum is the most common. Iron is found only in combination with other elements, which must be removed to

Table 5-4 Metal identification by chips.

	Copper	Brass and Bronze	Aluminum and Alloys	Monel Metal	Nickel	Lead
Appearance of chip	Smooth chips; saw edges where cut	Smooth chips; saw edges where cut	Smooth chips; saw edges where cut	Smooth edges	Smooth edges	Any shaped chip can be secured because of softness
Size of chip	Can be continuous if desired	Can be continuous if desired	Can be continuous if desired	Can be continuous if desired	Can be continuous if desired	Can be continuous if desired
Facility of chipping	Very easily cut	Easily cut; more brittle than copper	Very easily cut	Chips easily	Chips easily	Chips so easily it can be cut with penknife

	White Cast Iron	Gray Cast Iron	Malleable Iron	Wrought Iron	Low-Carbon Steel and Cast Steel	High-Carbon Steel
Appearance of chip	Small broken fragments	Small, partially broken chips but possible to chip a fairly smooth groove	Chips do not break short as in cast iron	Smooth edges where cut	Smooth edges where cut	Fine grain fracture; edges lighter in color than low-carbon steel
Size of chip		1/8 in.	1/4–3/8 in.		Can be continuous if desired	Can be continuous if desired
Facility of chipping	Brittleness prevents chipping a path with smooth sides	Not easy to chip because chips break off from base metal	Very tough, therefore harder to chip than cast iron	Soft and easily cut or chipped	Easily cut or chipped	Metal is usually very hard, but can be chipped

make it usable. It is widely used commercially for such things as bridge parts and castings. Wrought iron was the type of iron used before steel could be produced economically. Wrought iron is very low in carbon as a result of the way in which it is made.

Cast iron is wrought iron with the addition of at least 1.7 percent carbon and certain quantities of sulfur, phosphorus, manganese, and silicon. Gray cast iron is cast iron with the addition of graphite. *Graphite* is a form of free carbon that combines with pure iron during smelting. Its addition produces a form of iron that is softer than cast iron.

White cast iron is produced when scrap steel and pig iron are melted in an air furnace, or *cupola*. This special process produces a white cast iron that is very hard and brittle. Generally white cast iron is produced along the way to making malleable cast iron.

FIGURE 5.13

Use a bench grinder for the spark test.

When white cast iron is annealed, the end product is malleable cast iron. Annealing, basically, is a special heating process followed by a very slow cooling that creates an iron that is malleable. *Malleable cast iron* is the end product of a special process that heats up castings (made from white cast iron) and keeps them at a specific controlled temperature. Then the castings are cooled over a period of about 24 hours. Although the finished product is not by definition malleable, it possesses great toughness and is not brittle.

Steel

As you know from the discussion earlier, *steel* is an alloy of iron with a carbon content of no more than 1.7 percent. Steel is easily the most common metal for the welder, and there are many different types. By varying the content of carbon, or by adding other elements, different

1 Wrought iron

2 Mild steel

3 Tool steel

4 High carbon steel

5 High speed steel

6 Manganese iron or steel

7 Mushett steel

8 Magnet steel

FIGURE 5.14

Representations of the sparks obtained when various metals are lightly pressed against a grinding wheel.

types of steel can be produced. Steel with a low carbon content can be soft and easily worked by hand with light tools; steel with a high carbon content, however, will be very hard.

There is a very broad range of carbon steel in use today. Steel can be classified with either a number, such as 440 carbon steel, or simply by designations such as machine steel or tool steel. In any case, these designations will indicate the various qualities of a particular steel. For example, machine steel has a low carbon content, 0.10 percent to 0.20 percent, and is easily worked by a machine such as a lathe. Tool steels have a higher carbon content, at least 0.60 percent, but they lose their hardness at high temperatures because the carbon leaves the metal when heat is applied.

Carbon can be added to molten steel up to 0.83 percent. After this point, the tensile strength of the steel will be reduced. The higher the carbon content in the steel, the greater is the steel's resistance to wear. There is also a sacrifice as far as strength and toughness are concerned.

Alloy steels are created when other elements are added to plain carbon steel. Basically, an *alloy* is the combination of any two metals in any proportions. In the case of steel, there are many different elements and metals that can be added to produce a new steel with certain qualities. See Table 5-5 for alloying elements of steel.

Copper

Having been in use for over 4,000 years, copper has many applications around the home and in industry. Probably the greatest use is in plumbing pipes and water lines. Copper is very resistant to corrosion and oxidation. One of the unique properties of copper is that when it cools from a molten state, it expands. This is a favorable reaction when welding pipes and joints. When copper is heated to just below its melting point (about 2,000°F), it becomes as brittle as glass. Copper is a tough (tensile strength of about 20,000 psi) ductile metal that can be easily worked cold or hot. Copper is, of course, a nonferrous metal that produces little or no sparks and is nonmagnetic.

Aluminum

This metal is widely used in industry as well as around the home. Pure aluminum has a tensile strength of about 13,000 psi, which can be increased with alloying and heat treatment to about 80,000 psi. Aluminum can be worked by any method used commercially. Because aluminum is half the weight of iron, it is used extensively in the aircraft industry.

The surface of aluminum is oxidized when it comes in contact with the atmosphere, which can have an effect when welding. This semitransparent oxide film must be removed before welding can be successful. Aluminum poses special problems for the welder that are covered later in this book.

Table 5-5 **Principal alloying elements of steel.**

Element	Melting Point (°C)	Application	Result
Aluminum	658	Little aluminum remains in steel.	Deoxidizes and refines grain. Removes impurities.
Chromium	1615	Stainless steels, tools, and machine parts.	Improves hardness of the steel in small amounts.
Cobalt	1467	High-speed cutting tools.	Adds to cutting property of steel, especially at high temperatures.
Copper	1082	Sheet and plate materials.	Retards rust.
Lead	327	Machinery parts.	Lead and added tin form a rust resistant coating on steels.
Manganese	1245	Bucket teeth. Rails and switches.	Prevents hot shortness by combining with sulfur. Deoxidizes. Increases toughness and abrasion-resistance.
Molybdenum	2535	Machinery parts and tools.	Increases ductility, strength, and shock resistance.
Nickel	1452	Stainless steels. Acid-resistant tools and machinery parts.	In large amounts—resists heat, adds strength, toughness, and stiffness to steel.
Phosphorus (provided by ore)	43	Some low-alloy steels.	Up to 0.05 percent increases yield strength.
Silicon	1420	Precision castings.	Removes the gases from steel. Adds strength.
Sulphur	120	Some machined pieces.	Adds to the steel's machinability.
Tin	232	Cans and pans.	Forms a coating on steel for corrosion-resistance.
Titanium	1800	Used in low-alloy steels.	Cleans and forms carbide.
Tungsten	3400	For magnets and high-speed cutting tools	Helps steel retain hardness and toughness at high temperatures.
Vanadium	1780	Springs, tools, and machine parts.	Helps to increase strength and ductility.
Zinc	420	Wire, pails, and roofing.	Forms a corrosion-resistant coating on steel.
Zirconium	1850	Machine parts and tools.	Deoxidizes, removing oxygen and nitrogen. Creates a fine grain.

Magnesium

Magnesium is the lightest metal in use today. It equals about one-fifth the weight of copper and less than two-thirds the weight of aluminum. Although magnesium can be worked by all methods, it is subject to rapid oxidation. Therefore, special welding methods must be used. Magnesium is a metal that combines lightness in weight with excellent strength. Magnesium alloys can have a tensile strength of up to about 50,000 psi.

Sources of Metals

Before you can begin to learn how to braze, cut, or weld, you must lay your hands on a supply of metal. Probably the best place to start your search is in the Yellow Pages of the telephone directory. The categories to look under are steel, steel fabricators, steel products, metal, salvage, sheet metal, and metal stamping. You can also look under specific headings such as brass, iron, steel, aluminum, and stainless steel. You might also check under steel mills, if you know of any in your geographical area. Metal scrap yards are also a good source of relatively inexpensive steel and other types of metal. Your welding supply house, where you buy your oxygen and acetylene, should be able to tell you where you can purchase steel for welding projects. Local welders are another good source of information about where to buy metal for projects. In many cases, these people who make a living welding and repairing metal will often have scrap metal in limited quantities for sale. Professional welders are also a very good source of technical information and advice. It can be to your advantage to establish a relationship with a few.

Obtaining New Steel

In the beginning you should work with new steel. There will be fewer problems to deal with than if you begin working on rusted scrap steel, for example. However, obtaining new steel will mean a trip to a local steel fabricator. Here you should find a large selection of steel including sheets, plate, tubing, angle, rod, and strap steel. These companies operate by purchasing large quantities of steel from steel mills and keeping inventories for steel workers who must commonly buy in small amounts.

QUICK»»TIP When you pay a visit to your local steel fabricator or metal salvage yard, you should be prepared for the possibility of getting a little dirty from rummaging through scrap piles. Old clothes, boots, leather gloves, and a hat will do much to protect you from the generally dirty conditions usually found in these places. A small pocket ruler and a magnet are useful tools to take along as well. You should also be prepared to carry the metal home. Undoubtedly the best way is in the back of a pickup truck. If you don't own one, ask a friend who does to go along with you. As a last resort, you can pile the metal on top or inside your automobile. Roof racks and a rope will save a bit of wear and tear on the interior.

When you get the steel home, you will want to store it in such a way that it won't rust. If you don't have room inside, you will obviously have to store the metal outdoors. At the very least you should cover the steel with a plastic or canvas tarp.

It has always been my experience that the people who operate steel fabrication yards are quite cooperative. Most large companies will often be very willing to help you find the specific metal and gauge or thickness you need. If you are looking for a specific type of metal that is not generally stocked, a certain sheet steel, for example, a steel fabricator would be the place to start looking. Keep in mind that special orders for metal usually require a deposit.

Obtaining Scrap Metal

Once you have become familiar with the various brazing, welding, and cutting techniques using new steel, you will probably want to extend your basic knowledge to other metals—high-carbon and stainless steels and bronze. You may also want to try metal sculpture with either new or scrap metal.

Again, the steel fabricator will probably be your best source. It is a wise idea to obtain samples of many different types of metals such as chromium, molybdenum, and high-carbon steel so that you can widen your experience.

Scrap metal is usually very reasonably priced and commonly sold by the pound. Many interesting shapes, thicknesses, and types of both nonferrous and ferrous metals can be had with a single trip to the local metal salvage yard. Keep in mind that it will be necessary to devote more time to surface preparation when working with scrap metal. Many sculptors feel the extra effort is worthwhile.

Welding Supplies

Your local welding supply house is your best source for welding supplies and technical information. If you ever have a question about a specific welding project, this is the place to ask. Many of these supply houses also offer instruction and other services. Your welding supply house will be your source for oxygen and acetylene. You should also find a wide selection of tools, accessories, fluxes, welding, brazing, and special-purpose rods. This chapter will focus on rods and fluxes.

Welding Rods

During the welding process, it is often necessary to add more metal to the molten area to make the weld stronger and to fill in the area. This is accomplished through the use of a special *welding rod*. It is very important that the welding rod being used matches as closely as possible the metal being welded, the parent metal, for a strong fusion of the metal. During welding, the rod melts and deposits its metal into the molten parent metal and contributes to the strength of the weld. If the welder were to choose the wrong type of welding rod, the end result may be a weak and ineffective welded joint.

As you might expect, welding rods are available in a wide variety of sizes and compositions. There is easily a welding rod for every type of parent metal. The standard sizes of welding rods are 36

FIGURE 6.1

An assortment of welding and brazing rods. The square rod on the right is for cast iron welding.

FIGURE 6.2

Cast iron welding rods are square rather than round.

inches long for all types except cast iron rod, which is sold in 24-inch lengths. Diameters of gas welding rods range from $\frac{1}{16}$ to $\frac{3}{8}$ inch (Fig. 6.1).

The most commonly used gas welding rods are mild steel, alloy steel, and cast iron. *Mild-steel welding rods* are most commonly covered with a copper coating that protects the rod from rust during storage. Generally speaking, mild-steel welding rods are low in price and have a wide application for welding all types of mild steel with a low carbon content. The tensile strength of mild-steel welding rods is around 52,000 psi.

Alloy steel welding rods (for oxyacetylene welding) are used for relatively low carbon alloy steels. These rods are general-purpose welding rods that are frequently used for pressure systems such as pipelines. The tensile strength of these rods is about 62,000 psi.

Cast iron welding rods are available in several different types, depending of course on the application. For example, cast iron rods are available for welding gray cast iron, molybdenum cast iron (an alloy of cast iron), and straight cast iron. As mentioned, cast iron rods are available in 24-inch lengths. They are also most commonly square rather than cylindrical (Fig. 6.2).

Brazing Rods

Brazing rods are used when two metals will be joined by the brazing method. Basically, brazing rods are either bare or flux-coated bronze or brass rods that are used for brazing or braze welding steel, cast iron, brass, and bronze. Standard bronze brazing rods have a tensile strength (when used on steel) of about 50,000 psi. The melting point of standard bronze brazing rods is much less than that of other metals, around 1,600°F.

In addition to the popular bronze brazing rods, other types are also available for special metal joining situations. These special rods include those made from manganese-bronze, deoxidized copper, and nickel silver. Brazing and all types of rods for this metal joining process are covered in detail in Chapter 8.

> **QUICK>>>TIP**
>
> In addition to the three basic types of welding rods, there are many other special-purpose rods. These rods are used when joining other types of metals such as stainless steel and aluminum. Inquire at your local welding supply house for information about the various types of welding rods. You might also want to pick up an assortment of welding rods for experiments and practice.

Surfacing Rods

Still another type of rod used by the welder is a *hard-surfacing rod.* Briefly, certain tools and machinery that are commonly subjected to abrasion, such as a plow blade or the blade on a bulldozer, should be protected against wear by hardening or coating the edge of the tool that receives the punishment. One way to do this is to heat the forward area with a welding torch and then melt a special metal over it (Fig. 6.3). This additional new material will be extremely resistant to wear and will help prolong the life of the tool. Generally speaking, the most common hard-surfacing materials are nonferrous cobalt chromium tungsten alloys. These include nickel base, iron base, and tungsten carbide materials.

FIGURE 6.3

Application of a melted metal to the underside of a plowshare.

Fluxes

The success of metal joining, brazing, or welding is directly related to how clean the metals are prior to joining. In some metals such as aluminum, the formation of oxides during the joining process presents a special problem. These oxides must be removed from the surface being

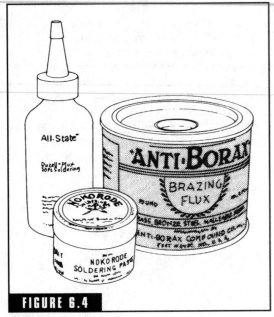

FIGURE 6.4

Fluxes aid in joining certain types of metals and are often required for soldering and brazing.

welded or brazed to insure a solid bond. In Chapter 4 we briefly discussed shielded arc welding (TIG and MIG welding) and how an inert gas is sprayed over an area during the welding process to prevent contamination. During oxyacetylene welding, there is no such protection, so another means must be used. This protection is accomplished through the use of special fluxes (Fig. 6.4).

A *flux,* according to the American Welding Society, is a material used to prevent, dissolve, or facilitate the removal of oxides and other undesirable substances from the area being brazed or welded. All fluxes are nonmetallic and fusible. Fluxes cause a chemical reaction to take place resulting in a slag that floats to the top of molten metal during welding or brazing. As you may have guessed, in order for this chemical reaction to take place the proper flux material must be used. There are fluxes for different types of applications. When welding some types of steel, flux is generally not required because the melting point of the steel is about the same as that of the oxides. Other metals, such as aluminum and iron, however, do not possess these qualities. Therefore, a flux must be used to lower the melting point of certain oxides so that they can be removed to ensure that a strong joint results. Special fluxes are always required when brazing metal.

Fluxes are available for brazing, welding, and soldering. They are sold as powder, paste, and liquid. Flux-coated brazing and welding rods are also available for special applications. It is always a good idea to check at your local welding supply house whenever you are in doubt as to the right type of flux to use for a particular welding project.

Soldering

here are three ways of joining metal: soldering, brazing, and welding. Of these, the strongest, of course, is welding wherein the
edges of two or more pieces of metal are heated up to the melting
point and they fuse together. The brazing process is similar except
that the metals are not brought up to the molten state, but a metal
filler material is. Soldering metals produces the weakest joint of all
three methods. In many cases, such as joining plumbing connections, strength is not as important as a complete sealing of the joint
surfaces (Fig. 7.1).

The principle of metal *soldering* is really quite simple. The pieces
are joined without bringing either up to the molten stage. Instead, a
special metal, solder, is used to bridge the gap between the two pieces.
When soldering can be used, it is quite effective. Technically the
process is called *adhesion*. In fact, what takes place is the molecules of
solder mingle with the parent metal's molecules and form a strong
bond. In some cases the ingredients in the solder, nonferrous metals,
form a surface alloy on one or both of the metals being joined. This
forms a solid bond between the metals.

Soldering is similar to brazing in that a nonferrous metal is used to
bridge the space between the two metals. The major difference
between these two methods, however, is the melting temperature of

FIGURE 7.1

Soldered joints are never very strong, but in some cases strength is less important than a good seal.

FIGURE 7.2

A typical soldering iron consists of a wooden handle, steel shaft, and a tip.

Table 7-1
Temperature Ranges, °F

Welding	2,800 °F
Brazing	2,000
Soldering	800

the joining nonferrous metals. All soldering is done at a temperature below 800°F. To give you an idea of the temperatures at which soldering, brazing, and welding are accomplished, please refer to Table 7-1.

Tools

Since not as much heat is needed for the soldering process, different tools are used than for brazing or welding. Soldering can be accomplished with any number of low heat generating tools such as a soldering iron, soldering gun, or any of the gas (acetylene, propane, natural gas, mapp gas) torches specifically designed for the purpose. It will be helpful to discuss each of these soldering tools and explain their possible uses.

Soldering Irons

Soldering irons are probably the simplest type of soldering tool. A typical soldering iron is simply a piece of steel rod with a wooden handle on one end and a large copper tip on the other end. Sizes can range from the small pencil-like lightweight types for small electrical soldering projects to large two-handed units used for special industrial applications. Any type of soldering iron must be heated up to a temperature hot enough to melt solder—up to 800°F (Fig. 7.2). This is generally accomplished through the use of a special furnace that is commonly gas operated, although some furnaces are electrically heated (Fig. 7.3).

There are other types of soldering irons that are much more portable because they are heated internally, either electrically or with some type of bottled fuel. Both of these types of soldering irons are illustrated in Figs. 7.4 and 7.5. The electrically heated soldering iron is best suited for continuous delicate work, while the gas-powered unit may have limitations because of the resultant flame below the tip.

Soldering Guns

Soldering guns are quite popular with do-it-yourselfers. They are fast, quick heating, and quite suitable for small repairs and electronic repair or building projects. There are a number of these soldering guns on the market, usually in some form of soldering kit that contains an assortment of tips for different types of soldering (Figs. 7.6 and 7.7).

Torches

Still another tool used for soldering work is a gas-powered, hand-held torch. There are several variations of this type, designed for specific applications. The standard do-it-yourself model has many uses around the home. The flame itself is used for heating the metal and melting the solder. Needless to say, a flame-powered torch can be used only for certain types of soldering (Fig. 7.9). A propane torch is the tool of choice (by professionals and do-it-yourselfers alike) for soldering the joints in a copper pipe plumbing system. The torch can also be used for thawing frozen

FIGURE 7.3

Old soldering irons, since they had no power of their own, had to be heated in a special furnace that was most commonly gas fired.

FIGURE 7.4

This soldering iron tip is heated by the flame from a propane torch.

FIGURE 7.5

An electrically heated soldering iron is very handy for continuous soldering.

FIGURE 7.6

A soldering gun kit contains several different size tips and is handy for a wide range of soldering projects.

FIGURE 7.7

A hand-held soldering gun is useful for many soldering projects.

water lines, removing paint, and even heating vinyl asbestos floor tile prior to installation. The most common fuel for this type of torch is *propane,* sold in pressurized cylinders that contain enough propane to power the torch for about 2 hours. Mapp gas is another bottled fuel that can be used in a hand-held torch. *Mapp gas* is available in standard size cylinders (commonly yellow in color) and offers a much hotter flame than conventional propane.

Propane-powered torch kits are available and are a good investment for the do-it-yourselfer. These kits commonly contain a variety of attachments such as a soldering tip and various size burner heads (Fig. 7.8).

Professionals who must solder many joints in the course of a day's work will usually have a special torch for this purpose. These torches, popular with plumbers, have a hand-held torch unit connected by a pressure hose to a regulator that fits into a valve on the top of a 25-pound tank (or larger) of acetylene. These torches can also be run on propane or LP gas and are truly a time-saving tool when used on a daily basis. This type of low-heat torch outfit is also standard equipment in any radiator repair shop (Fig. 7.10).

Copper Tips

All soldering tools with *copper tips*—excluding hand-held torches without tips—should provide years of dependable service with little care. Undoubtedly the most important part of any soldering tool is the tip of the unit. This part is the only one that comes in contact with the work. To ensure that this important part is able to do its intended task of heating the metal and melting the solder, there are a few things you must do to it. We will discuss these maintenance procedures in detail.

The tip of a soldering tool should never be allowed to overheat. On electrically powered units, this is rarely a problem as the electrical current is controlled. But soldering irons that must be heated in a furnace are often prone to overheating. Overheating of the tip will result in excessive scaling of the soldering iron core and carbonizing or rosin fluxes. It is difficult to properly tin an overheated soldering iron tip. One indication of overheating of the soldering iron tip is a greenish tinge that will appear when the tip is hotter than it should be.

A good propane torch kit will contain a variety of useful attachments.

The Spitfire brazing torch has an attachment that aids in soldering plumbing pipe.

Fluxes, which we will discuss in greater detail later, can have a corrosive effect on soldering tips. While fluxes are necessary, you should use the mildest flux possible for the particular job at hand. When you have completed a soldering project, you should clean the tip of flux residue with a damp cloth or steel wool.

If your soldering tool is electrically operated, you should unplug it when not in use. This is particularly important for those soldering tools that remain hot as long as they are plugged in. For soldering guns, with a trigger that activates the heat source, this is not quite as important.

While using the soldering tool, you will achieve better results if you keep the tip well coated with molten solder. This is commonly called "tinning the tip," a procedure that will be covered later in this chapter.

Soldering tips will become dirty in time through normal use. For this reason all tips should be cleaned periodically. In most cases a cleaning with a wire brush or steel wool will restore the tip to a like-new condition. Care must be exercised when doing this, especially in the case of copper-coated tips, as the coating can be worn off if cleaned with a tool that is too abrasive. A grinding wheel should never be used for cleaning a tip, for example. After the tip has been cleaned, it should be immediately retinned.

FIGURE 7.10

A portable torch outfit is suitable for light cutting, welding, brazing, and soldering.

TRADE SECRET Most authorities recommend that a soldering iron tip never be brought up to a temperature in excess of 750°F. In most cases, soldering can be accomplished at lower temperatures anyway, so there is little point in bringing the iron up to this temperature.

Probably more damage is done to soldering tools through rough handling and dropping than through normal wear and tear. It will be to your advantage to use your soldering tool properly, according to the directions of the manufacturer. Be careful when using a soldering tool because the possibility of burning yourself always exists. A soldering iron holder is good insurance against accidental burns and also serves to keep the tool handy when not in use. You can usually fashion a soldering tool holder out of a piece of steel welding rod or even a wire coat hanger (Fig. 7.11). You can also rest the hot iron on a brick or other suitable surface.

FIGURE 7.11

A soldering iron holder can be made easily from a piece of welding rod or coat hanger.

Choosing the Proper Solder

The key to successful soldering lies in choosing the proper solder for the project at hand. As you might expect, there is quite a large selection of different types of solders currently on the market. The two most common types of solder used outside the industry are those designated as *50-50* and *60-40*. These two are the easiest for the do-it-yourselfer to use. Type 50-50 solder is 50 percent tin and 50 percent lead, while 60-40 solder is 60 percent tin and 40 percent lead The percentage of tin is always listed first. A chart of the six most commonly used solders is shown in Table 7-2. All solders except the high-temperature, 5-95 solder, melt at a rather low temperature. You will also notice, however, that the more lead in the solder, the higher the temperature must be before the solder will flow or become liquid. This is due largely to the fact that the greater amount of tin in a nonferrous alloy, the lower the flowing temperature. The addition of tin lowers the alloy's melting temperature. In an effort to give you a better understanding of these six common solders, I would like to briefly explain the uses of each (Fig. 7.12).

5A SOLDER

This solder has greater strength quality than the other types because of the high lead content. It is a good solder for torch soldering where the flame is used rather than a heated tip. This solder can be difficult to work with unless you are quite experienced. For this reason it is generally a poor choice for the do-it-yourselfer. Nevertheless, 5A solder can provide a very strong joint.

FIGURE 7.12

Most repairs on automotive radiators are made with solder.

Table 7-2 Tin-Lead Solders

Number	ASTM % Tin	% Lead	Melting Temperature	Flowing Temperature
5 A	5	95	572 °F	596 °F
20 A	20	80	361	535
30 A	30	70	361	491
40 A	40	60	361	455
50 A	50	50	361	421
60 A	60	40	361	374

20A SOLDER

This type is used extensively in certain industrial soldering tasks such as sealing automobile radiators and auto body work (Fig. 7.13). This is a good solder to use where the finished joint or joints will not be exposed to high temperatures that would cause joint deterioration.

30A SOLDER

It is similar to 20A solder in that it is widely used for industrial work. This is a common solder for automobile body repair and torch soldering of all types.

40A AND 50A SOLDER

These are probably the most widely used general-purpose solders available today. Both types are used for plumbing pipe

FIGURE 7.13

One of the best solders for repairing a radiator is 20A.

joints, roofing seams (where tin or copper is the covering material), and for electrical circuit connections such as radio and television. When used for electrical work, the solder usually has a rosin core instead of a solid core. A rosin core solder contains a noncorrosive flux that is made from pine trees (Fig. 7.14).

60A SOLDER

This is often called *fine solder,* and it is used wherever temperature requirements are critical, such as in precision instruments.

Another group of solders worth mentioning is called *silver solder.* This group can contain either lead and silver, tin and silver, or a combination of all three nonferrous metals. Silver solders are used for special joining tasks, such as for certain types of metal or where electrical conductivity is important, particularly radio or other electrical component circuitry. Silver soldering produces a strong joint. Generally higher soldering temperatures and special fluxing techniques are called for, not to mention greater skill during soldering.

FIGURE 7.14

General-purpose solders include 40A and 50A.

Leadless Solder

There has been growing concern about the amount of lead we are all exposed to on a daily basis, and a number of regulations have been put in place to reduce how much lead we come in contact with. Consider leadless paint, unleaded fuel, and steel shot.

When copper pipes are soldered—with lead solder—minute traces of lead can get into our drinking water. It is probably safe to say that most of the homes in America have plumbing systems that were build with lead solder. Even small amounts of lead in drinking water can cause headaches, bone and teeth deterioration, brain damage, and in excessive amounts, death. The very young and the elderly are particularly susceptible to the effects of lead.

In recent years an alternative to conventional lead solder has been developed— leadless solder. This new product is widely available in home improvement

QUICK TIP Because of its composition, leadless solder behaves slightly differently from traditional lead solder, and many professional plumbers prefer not to use it. The melting point of leadless solder is higher than lead solder, and more heat must therefore be applied before it will flow. Once you have worked with leadless solder for a short time, you should be able to sweat solder copper plumbing pipe as easily as with leaded solder, however.

centers and plumbing supply houses. In use, leadless solder is applied to sweat solder joints in the same fashion as leaded solder. The main advantage of leadless solder is, of course, that no lead can get into your drinking water.

Fluxes

One of the requirements of successful soldering is that the metal surfaces be chemically cleaned. They must remain in this state until the solder has hardened (Fig. 7.15). Flux was designed for this purpose. When properly used, it not only cleans the surfaces but seals them from the atmosphere during the soldering. Additionally, some fluxes add alloying elements to the solder or parent metal while other fluxes reduce the surface tension on the metal, which increases the fluidity and adhesion of the solder.

At this time, there is no universal flux. Instead, there are several different types of fluxes designed for certain types of metals. Fluxes are available in powder, paste, and liquid form.

Noncorrosive Fluxes

All fluxes can be grouped into either the noncorrosive or corrosive category. Generally speaking, *noncorrosive fluxes* are mild or weak in their fluxing action. These types are recommended for all electronic and other critical soldering projects. Noncorrosive fluxes are generally suitable for clean copper, brass, bronze, tinplate, cadmium, nickel, and silver. Noncorrosive fluxes leave an unsightly brown stain on the surface. Since this residue is not electrically conductive, it must be removed after soldering.

Corrosive Fluxes

Corrosive fluxes are composed of inorganic ingredients. These fluxes are used when a rapid, highly activated fluxing

FIGURE 7.15

Fluxes and solders go hand in hand.

action is desirable, such as when soldering cast iron, sheet metal, copper, and brass. Corrosive fluxes generally work well with any of the methods commonly used for soldering.

Corrosive fluxes leave a residue after soldering that can chemically attack the joint just soldered as well as the surrounding metal areas. For this reason, the residue from these fluxes must be removed after soldering. Corrosive fluxes are unsuitable for certain types of soldering projects such as sealed containers and most electrical connections because of the corrosive action of the flux. One of the advantages of corrosive fluxes is that they remain relatively stable over a wide range of temperatures, which makes them suitable for soldering projects requiring higher temperatures.

Probably the most common type of corrosive flux is called *acid flux* (Fig. 7.16). A common formula for this flux consists of 70 percent zinc chloride and 30 percent ammonium chloride. This type is called "acid flux" because as the flux is heated and melts, it forms hydrochloric acid. The acid dissolves the oxides and permits the solder metal to adhere to the parent metal. As mentioned, corrosive flux residue must be removed after the soldered joint has hardened (Fig. 7.17).

FIGURE 7.16

An acid flux.

FIGURE 7.17

Corrosive flux should be removed after soldering. A damp cloth will often do the trick. If not, baking soda can be tried.

Flux and Solder Combinations

Table 7-3 lists some of the possible flux and solder combinations. Keep in mind that in some cases other solder can be substituted, such as 50-50 solder for 60-40. But the best general results will be achieved with the combinations listed. It should also be noted that fluxes cannot generally be interchanged. If a noncorrosive flux is called for and a corrosive flux is used, such as in the case of electrical

Table 7-3   Suggested flux solder combinations.

Metal	Type*	Flux Solder	Flow Temperature
Tin	N/C	50−50	421 °F
Stainless steel	C	50−50	421
Copper and alloys	C	60−40	374
Iron	C	60−40	374
Galvanized metals	C	50−50	421
Steel	C	40−60	455
Electrical connections	N/C	40−60	455

* N/C Noncorrosive, C Corrosive

wiring, damage may be the outcome of the project. On the other hand, if a corrosive flux is called for and a noncorrosive flux is used, complete bonding may not happen because the flux used was not strong enough to do the job.

Surface Preparation

The first rule of successful soldering is cleanliness. Everything that comes in contact with the solder—the metals being joined, soldering iron tool tip, and even the solder itself—must be free of dirt, grease, oil, paint, rust, or corrosion on the surface—in short anything that may interfere with the solder's adhesion ability should be eliminated.

Of course, the type of project you are soldering will dictate the type of surface preparation. Some of the materials available for cleaning metal surfaces include steel wool, sandpaper, files, disk sanders, grinding wheels, and wire brushes (Fig. 7.18). You should use any practical method to bring the surfaces up to a bright and shiny condition. Just before you begin soldering, it is also a sound practice to lightly polish the solder itself with either steel wool or sandpaper. This simple act will remove anything on the surface of the solder and ensure that you are using pure solder (Fig. 7.19).

A second rule for successful soldering is that the pieces must fit together well before the actual soldering begins. Don't expect the solder

FIGURE 7.18

Some of the common tools used in surface preparation.

to stand up under a lot of stress and strain because it generally won't. It is far better to have a tight mechanical bond before soldering the work together than it is to expect the solder to do this for you.

In some cases achieving a sound connection or fit will be fairly simple, as in fitting plumbing connections together. But, in other cases, you will have to do quite a bit of bending and hammering to get the pieces to fit together properly before soldering. In all cases, a gap of about 0.002 inch is acceptable. If the gap is any larger, the solder may not have the ability to hold the work together. The result may be a failing joint.

Whenever possible, you should avoid butt joints, as these are the weakest type for soldering. Figure 7.20 illustrates some of the more common metalworking joints that can be used when soldering. As you can see, most of the joints are of the interlocking type. Fashioning this type of joint requires skill that can be developed only through familiarity with the metals being worked and with time. Nevertheless, there are a few tools that can help you to make some of these joints. These tools include a

FIGURE 7.19

Clean solder with steel wool before using. This will remove surface dirt.

Single edge

Double edge

Wired edge

Flat band

Plain lap

Strapped butt joint

Riveted lap

O-geed

Curb angle

Flush lock seam

Flanged riveted

Keyed lock

Plain lock

Side lock seam

Standing end lock

Folded end lock

Lead joints

Cast brass fittings

Inside lock

Lock seam

Wrought copper fittings

FIGURE 7.20

Soldering joints in common use.

FIGURE 7.21

A pair of pliers is very handy for bending sheet metal to form a soldering joint.

FIGURE 7.22

Vise-Grips are a handy tool to hold pieces secure for soldering.

pair of pliers, a bench vise, hammer and anvil, and your own ingenuity. Remember that the better the metals fit together, the better the finished joint will be (Fig. 7.21).

Keep in mind, when fashioning joints from sheet metal, that this was the way many metals were joined before welding came into existence around the turn of the century. Forming solid joints prior to soldering is good practice for similar work with metals, and it is covered in later chapters. In soldering, the better the pieces fit together, the stronger the joints and the better the finished project will be.

Technique

Once the proper fit or joint has been established, the pieces can be soldered. Before actually beginning, however, it may be necessary to hold the pieces together with a clamp or other suitable tool. This will keep the pieces aligned while you are working on them. It will also be good insurance that the pieces are not disturbed while the solder is cooling, which could result in a weak or ineffective joint (Fig. 7.22).

Tinning the Tip

If you will be soldering with an iron or soldering gun, the first task you must accomplish before actually soldering is to heat up the tool and tin the tip. *Tinning the tip of a soldering tool* simply means coating the tip with molten solder. Since it is impossible to solder with an untinned or badly tinned tip—the oxidized film of copper on the surface of the tip prevents the ready transmission of heat—it is important to tin the tip thoroughly. Actually, tinning the tip is a simple process. Begin by heating the soldering tool until it will melt solder (Fig. 7.23). Then simply rub some solder onto the hot tip, and spread the molten metal over the surface. When this is properly done, the tip will have a bright, shiny silver appearance.

If you have a problem getting the solder to stick to the tip, it may be because the tip of the tool is dirty. Clean the hot tip with steel wool or a damp cloth; then try again (Fig. 7.24). If the tip is too hot, the copper will tarnish immediately. When this happens, it will be necessary for you to let the tip cool slightly before tinning. You may also find that a little flux rubbed on the tip before applying the solder will enable you to tin the tip a bit more easily (Fig. 7.25). Use either corrosive or noncorrosive flux, depending on the project at hand.

FIGURE 7.25

To tin the tip of a soldering iron, begin by heating the tip and then simply melt solder to cover it.

Heating the Work Metal

After the tip of the soldering tool has been sufficiently tinned, you can begin the actual soldering. Begin by first applying some flux to the area being soldered. The proper way to solder is to heat up the work with the soldering tool and then apply the solder to the work. The desired technique is to let the heat of the metal melt the solder rather than to melt the solder with the tip of the soldering tool. The obvious reason for this approach is that when the

FIGURE 7.24

Clean the copper tip of a soldering iron with steel wool before tinning with solder.

metal is hot enough to melt the solder, you can then expect a reasonably good bond. When the metal is not at the proper temperature, all the solder will do is lie on the surface rather than be bonded to the surface (Fig. 7.26).

There are a number of factors that will determine how quickly you can heat up the work metal that is to be soldered. These include the thickness of the metal, the size of the tip of the soldering tool—larger tips heat a larger area than small soldering tips—and the power or heating ability of the soldering tool. For small projects that may

FIGURE 7.25

Sometimes it is necessary to apply flux to the soldering iron tip before tinning.

FIGURE 7.26

Always heat up the metal with the soldering iron; then apply the solder to the hot joint. Never do it the other way around.

require only a spot of solder, such as electrical connections, you should use a very small tipped soldering tool. For large soldering projects such as a seam in sheet metal, you will find the work much easier to accomplish if you use a soldering tool with a large tip (Fig. 7.27).

Tinning the Surface

Once the metal is hot enough to melt the solder, your first objective is to tin the surface. The word *tinning,* in this case, is derived from an old sheet metal working term. It means to apply a very thin layer or film of solder to the metal surface. Your aim is to lay a thin layer along the entire joint or seam, which will cause the pieces to bond or stick together. Then go back over the work and build up the bead of solder until you have a solid, strong, and even seam.

In some cases, such as installing a radiator bracket, both surfaces are tinned before assembly. Then the pieces are pressed together and heated. This causes the mating surfaces, which have already been tinned, to adhere to one another (Fig. 7.28).

As I mentioned earlier, the solder will always flow toward the heat source. You can take advantage of this fact when you need to make solder slow uphill, as when joining copper tubing for an overhead plumbing system or when soldering a vertical seam. Move the heat source slowly along the seam, and the molten solder will follow, leaving a solidified trail.

Chunky Solder

When soldering is being done properly, the cooling solder will appear smooth and slightly shiny. If you are moving too quickly, the solder

FIGURE 7.27

Small soldering projects are easier to accomplish with a small size soldering iron.

may take on a grayish cast. This means that the metal was not heated quite enough to flow the solder totally. In extreme cases, the solder will appear *chunky,* with little sharp-edged bits of solder in the bead. This is a problem that indicates you are working too quickly and the solder may be the type that requires more heat to flow properly. For example, 5-95 or 20-80 solder requires higher temperatures than 50-50 solder (Fig. 7.29).

Cooling and Cleaning the Soldered Area

After the joint or seam has been soldered to your satisfaction, remove the soldering tool. Remember that the solder will follow the heat. Once you have removed the heat source, look over the work carefully. If any sharp points are present, or any areas appear not to be covered sufficiently, touch up with the tip of the soldering tool as necessary. When you are satisfied with the joint or seam, let the work sit undisturbed until it has cooled because moving the still hot metal may cause the seam to come apart.

Most soldering that involves the use of flux will also require that you clean the just soldered area after it has cooled. In the case of noncorrosive

FIGURE 7.28

Some joints are easier to solder if the mating surfaces are tinned with solder and then fitted together and heated until the solder, on both faces, melts together.

FIGURE 7.29

A solder joint that appears "chunky" usually has not been preheated sufficiently.

flux, there is no danger that the flux will damage the joint or surrounding area. There will, however, be a brown residue around the soldered area, especially when rosin core solder was used. This brown film is very brittle—it is actually a pine tar residue—and it can be removed quite simply with the tip of a pocket knife or other suitable tool (Fig. 7.30).

Corrosive flux must be removed from the finished soldering project for reasons that were stated earlier—the potential of the corrosive flux to damage the surrounding metal. This is usually best accomplished by rinsing the area with water. If this is not possible, use a damp cloth (Fig. 7.31).

Soldering with a Flame

When soldering with a flame, instead of soldering with a copper tip, the same basic procedures are followed:

- Clean the surfaces.

- Fit the pieces together.

- Apply the flux.

- Heat the metal until it is hot enough to melt the solder.

- Let the metal cool undisturbed.

- Clean the soldered surface of flux and smooth it if necessary.

Generally speaking, soldering with a torch requires a light touch with the flame and strict attention to the work. Since the flame can be hotter than 800°F, the possibility always exists of overheating the metal. This can result in poor adhesion of the solder, scorching the metal, and, in extreme cases, warping the metal.

Soldering Copper Plumbing Pipe

Probably the most common use of a torch in soldering is for making soldered joints in a copper pipe water line system. It will therefore be helpful to explain how this is done.

Begin by cleaning the end section of pipe with either fine emery cloth or steel wool (Fig. 7.32). The same treatment is given to the inside of the copper fitting. All mating surfaces should be bright and shiny. Next, apply the flux (Fig. 7.33). In most cases this will be a corrosive paste, a standard in the plumbing trade. The pieces should then be fitted together tightly and positioned so that they will not move during the soldering process or until the solder has cooled and solidified totally.

Apply the heat to the pipe and fitting, moving the torch to ensure that the parts are being evenly heated. As the flux heats up, it will begin to bubble and possibly turn a brownish color. This is usually an indication that the copper is at about the right temperature to melt the solder. Move the flame of the torch away from the joint and touch the end of a strip of solder to the joint area. If the metal is hot enough, the solder will flow into the joint (Figs. 7.34 and 7.35).

FIGURE 7.30

You can usually remove noncorrosive flux with the tip of a pocket knife.

FIGURE 7.31

Remove corrosive flux from a cooled soldered joint with a damp cloth.

When soldering copper water lines, it is important that you do not apply too much solder. Actually you do not need very much solder. Only a small amount of solder should be visible around the outside of the joint. Excess solder buildup around the fitting and pipe may actually weaken the joint rather than make it stronger.

After the proper amount of solder has been applied to the joint, wipe the area with a damp cloth. This will give the joint a nice finished

FIGURE 7.32

Both the end of the copper pipe and fitting must be clean and shiny. They can be cleaned with steel wool.

FIGURE 7.33

Flux can be applied to the clean pipe and fitting with a small brush.

appearance, as well as clean the area of flux. The wiping will also remove excess solder (Fig. 7.36).

Safety Considerations

Soldering is a relatively safe method of joining metals, but, as with any operation that involves heat, there is always the possibility of burns, noxious fumes, and damage to the skin. In most cases, accidents can be avoided by being careful while working. It is a good idea to have a specific place for the soldering tool when it is turned on but not actually in use. This can be a wire soldering iron holder as shown in Fig. 7.11, a large ashtray (although these tend to have a rather short life in the workshop), or even a brick. If the soldering iron or gun is electrically operated, as most types are, you should set up your work area so that there is not any possibility of burning the power cord with the tip of the unit. This could cause a short circuit or even an electrical shock.

Eye Protection

When working with corrosive fluxes, it is a good idea to wear a long-sleeved shirt and eye protection. In some cases a pair of gloves may also be a good idea.

The chances of mishap increase when you find that you must work away from a workbench. This is usually the case when working on a copper water line system. Precautions should be taken. If working overhead, wear eye protection, a long-sleeved shirt, and a hat.

Adequate Ventilation

Whenever you are soldering with a gas-powered torch like propane, you should make certain that there is adequate ventilation in the work area. If you are soldering copper pipes, keep an eye on combustible areas around the work—such as wooden flooring, joists, and wall studs. In many cases, you will be forced to solder very close to these areas. Know that the possibility of fire is always present, and have some means of putting out a fire if one should start. Some good fire protection or fighting tools to have around are a bucket of water, fire extinguisher, a box of baking soda, or a plant atomizer. The last item will quickly spray a cloud of water on a smoldering or burning piece of lumber and quickly put out the flame.

If you find it necessary to solder close to wooden joists, slip a piece of asbestos board between the work and the joist. The asbestos will prevent the possibility of a fire.

Since all fluxes give off fumes or smoke when heat is applied, ventilation is more than important. While the noncorrosive rosin fluxes are not harmful, fumes may upset your system for a short period of time. It is therefore not a good idea to breathe these fumes for long periods. Prolonged inhalation of corrosive flux fumes should be avoided. If the soldering project is a large one, some type of respirator should be worn to protect the worker.

Soldering is an effective and enjoyable means of joining metal that can be accomplished rather inexpensively. It is good practice for the

FIGURE 7.34

The fluxed fitting is twisted onto the pipe; then heat is applied with a propane torch.

FIGURE 7.35

After the pipe and fitting have been heated to the proper temperature, solder is touched to the hot joint and it will flow into place, forming a tight seal.

stronger methods of joining metals since some of the techniques are basically the same. Approach all soldering tasks with a bit of caution, and develop safe work habits. Not only will this help you to achieve predictable results but it will also help prepare you for other methods of joining metal.

FIGURE 7.36

Before the solder has totally hardened, wipe the joint with a damp cloth. Work carefully or you may disturb the joint.

Brazing and Braze Welding

In the previous chapter we covered how to join metals by using low heat and a nonferrous filler material. You will recall that soldering takes place at temperatures below 800°F and produces the weakest joint of the three types of metal joining techniques: soldering, brazing, and welding. In this chapter we will cover how to join metals, again by using a nonferrous filler material, only in this case at temperatures above 800°F. First, it will be helpful to us in our discussion of metal joining to explain and define the terms *brazing* and *braze welding* because they are often confused.

Uses of Brazing

Brazing and *braze welding* describe two different techniques for joining metals. *Brazing* refers to a process whereby the metal parts are heated to temperatures above 800°F (Fig. 8.1). Then a nonferrous filler metal is introduced into the joint. Part of the key to successful brazing is that the filler metal must have a melting point lower than that of the parent metals. The pieces to be joined are heated up to a point where the brazing filler metal will melt but not the parent metals.

When brazing metals, the fit between the pieces is very important and must be tight. Since the filler material flows freely over the surface and into the recesses of the joint, it will not fill gaps or other areas that

FIGURE 8.1

Brazing in progress.

do not fit together well. This flowing of the filler material is known as *capillary attraction.* It happens when the contacting surfaces of a liquid (the molten filler material) and a solid (the parent metals) interact so that the liquid surface is no longer flat. A coating of flux aids the capillary attraction and helps the filler adhere to the parent metals.

A brazed joint is stronger than a soldered joint, and, in certain instances, it is as strong as a welded joint (where the parent metals have been heated up to the melting point and have fused together). Brazing is used where mechanical strength and pressure-proof joints are required, such as in a pipeline system and cast transmission cases (Fig. 8.2).

Generally speaking, brazing, because it can be accomplished at lower temperatures, will not warp the metals being worked on. Brazing is also much quicker and less expensive than welding because the metals need not be heated up to the melting point.

FIGURE 8.2

The Spitfire torch can be used for light brazing.

Brazing is a suitable means of joining many types of metals such as steel, cast iron, nickel, copper, brass, bronze, cast bronze, magnesium, aluminum, monel, and inconel. For success when joining these metals, you need to fit the pieces together carefully. Use the proper flux and filler material, and work at a predetermined temperature. Brazing is also a good technique for joining dissimilar metals such as nickel alloys and copper alloy base metals.

Uses of Braze Welding

Braze welding differs from brazing in the design of the types of joints being secured. Braze welding is used for groove, fillet, and other weld joints where the filler material does not take advantage of capillary action. Braze welding does, however, use a nonferrous filler metal that has a melting point below that of the parent metal but above 800°F.

The technique for braze welding is similar to that used for fusion welding. The important exception is that the parent metal is not melted but only raised to the tinning temperature.

Filler Material

Both brazing and braze welding use a nonferrous filler material. In most cases this will be an alloy of copper, zinc, and copper or tin.

The two most commonly used copper alloys for brazing and braze welding are brass and bronze. Brass is an alloy consisting of copper and zinc, while bronze is composed mainly of copper and tin. The more popular for both joining techniques is brass. It is slightly interesting to note that even experienced welders call brazing and braze welding rods "bronze" when actually they are using brass rods consisting of copper, zinc, and about 1 percent tin (Fig. 8.3).

There are other brazing and braze welding rods that are used for joining special metals in addition to the copper alloys. These include nickel and chromium alloys, silver alloys, copper and gold alloys, aluminum and silicon alloys, and magnesium alloys. It is probably a safe assumption that the do-it-yourself welder will use a bronze rod for most types of brazing and braze welding. This is a good nonferrous filler material for joining cast iron and steel. For joining other types of

1/4"

1/8"

1/16"

FIGURE 8.3

Three common sizes of bronze rod used for brazing.

metals, it is always a good idea to consult your local welding supply house to find out the best type of nonferrous filler material.

Flux

Flux plays a very important role in braze welding and brazing. Most brazing processes require the use of flux to prevent oxidation during the application of heat. A flux, you may remember, is used in soldering to protect the metal surfaces being joined from the atmosphere. In brazing and braze welding, the flux is used for the same reason. In addition, a flux will aid the capillary attraction of the nonferrous metal while brazing and at the same time aid in dissolving the oxides that form during either brazing or braze welding.

Flux is commonly available in powder, paste, and liquid forms. Also available are bronze rods that have been coated with a flux, which eliminates the need for additional flux. There is no such thing as an all-purpose flux. There are, instead, fluxes designed for specific applications. Most fluxes sold are chemically pure and will have some mention of this fact on the label.

Borax is a flux that has been used for many years with great success for brazing brass, bronze, steel, and malleable iron. One of the more popular mixtures of this flux consists of 75 percent borax and 25 percent boric acid in paste form. This "antiborax" flux, as it is sometimes called, is suitable for brazing those metals men-

QUICK TIP Since the fumes from most fluxes can be harmful to your health, it is important that good ventilation be present in the workshop area. Some fluxes, such as those containing sodium cyanide salts, are very dangerous. You should avoid breathing the fumes or even letting the flux come in contact with your skin. Always check the container label for specific cautions.

tioned earlier. Other fluxes should be used for different metals, however, since there is no universal flux.

Most fluxes are sold in a powder form. This is quite convenient as the flux can then be applied to the metal in two different ways. The first way is to heat up the end of a bronze rod and dip it into the powdered flux. The hot rod will cause the flux to adhere to the rod, and the rod can then be used as it is, with a coating of flux (Fig. 8.4). Powdered fluxes can also be made into paste fluxes, usually by adding water until the right consistency is achieved (Fig. 8.5). Paste fluxes are commonly applied to the metal with a small brush. It is important that you read the container label as some fluxes require that you mix them with alcohol instead of water to form a flux paste.

Preparing Surfaces

Easily 75 percent of the success of any brazing project is directly related to the condition of the surfaces being joined. Not only must the surfaces be physically and chemically clean, but the pieces must fit together precisely. A joint clearance of 0.002 to 0.006 inch is acceptable when brazing copper or silver alloy. When joining aluminum, greater tolerances are permissible, as much as 0.006 to 0.015 inch. As you can imagine, joint design is quite important, and it will be discussed in this chapter. The clearances of a joint will have a direct bearing on the strength of the finished project. There must be a certain amount of clearance so that the nonferrous

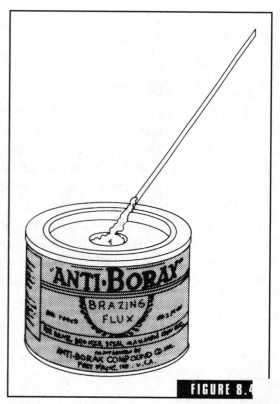

FIGURE 8.4

A hot brazing rod is dipped into the flux and is quickly coated.

FIGURE 8.5

Mix up water and powdered flux to form a paste.

metal filler can be drawn by capillary action into the joint. If the clearance is too great, however, the capillary attraction will be lessened. The end result will be a joint that is weak.

Cleaning Metal

There are several ways to prepare metal before the pieces are fitted together for brazing. These include cleaning them with wire brushes, emery cloth, crocus paper, or steel wool and, in special circumstances, washing the parts with an acid solution. If the metal has just been machined or milled, all that is usually required is a light touch with steel wool. But the older the metal, in relation to elements, the more surface preparation will be required.

If rust is in evidence, for example, you should begin cleaning with a wire brush (Fig. 8.6). In most cases this tool will remove all of the surface scale and oxidized metal. Next, the metal should be sanded with emery cloth or crocus paper to make the surface smooth and shiny.

It is generally not recommended that a grinding wheel or other harsh abrasive be used on metal that will be brazed. The main reason is that these tools have a tendency to impart scratches or a rough texture to the surface. Knowing that a tight joint between the metal parts is important, this type of cleaning may cause tolerances that are unacceptable, at least from the standpoint of brazing. A grinding wheel is suitable for surface preparation of metals that will be braze welded (Fig. 8.7).

If much scale is in evidence on steel or iron and it cannot be easily removed with wire brushes, steel wool, or emery cloth, it may be necessary to soak the pieces in a special *pickling bath*. This is simply a mixture of water and sulfuric acid. After the pieces have soaked in the bath, they must be rinsed in warm water until all of the acid residue is flushed away. Since working with acid can be dangerous, you should proceed with due caution. Your

FIGURE 8.6

A wire brush is very handy for removing surface scale and dirt.

local welding supply house should have the stock ingredients for a pickling bath, and they can probably offer some good suggestions as to how you might best go about this method of cleaning steel and iron. They may even perform this service for you. Generally speaking, a pickling bath will clean the metal better than any other method. It is a useful means of cleaning both small pieces as well as those metals that might require an enormous amount of work to clean.

FIGURE 8.7

While a grinding wheel is not generally suitable for metal that will be brazed, it can be useful for metal that will be braze welded.

Butt and Lap Joints

After the metal has been cleaned, the pieces must be fitted together so the tolerances are close, to within 0.002 to 0.006 inch. The best way to make certain that the pieces fit together this closely is to use or develop a joint that is tight fitting. The two most common joints used are the *lap* and the *butt joints*. Whenever possible, the former is much preferred.

The main reason that the lap joint is popular when brazing metals is that it offers the strongest joint. It is also relatively easy to make. Simply lap two surfaces together and braze. The length of the lap should be equal to at least three times the thickness of the thinnest member being joined. For example, if you were brazing two pieces of ³/₈-inch-thick steel, using the lap joint, the length of the lap should be at least 1¹/₈ inches. Figure 8.8 illustrates some of the more common joints, both good and bad.

Applying Flux

After the pieces have been cleaned thoroughly, the joint surfaces must be coated with the proper flux. Remember that there are different fluxes. You must choose the right one for the metal you are working with. In most cases a paste flux is brushed on the surfaces. If you are using a powder flux, follow the directions on the container to form a paste, and brush this on the joint surfaces (Fig. 8.9).

Once the pieces have been fluxed and fitted closely together, they must be held in this position until they are brazed and the filler material

FIGURE 8.8

The three most commonly used joints in brazing.

has cooled to a solid stage. In many cases a C clamp can be used for this. As an alternative, a pair of locking jaw pliers can be quite handy. Locking jaw pliers come in many different sizes, with ends in many different shapes. Your local welding supply house will probably have a selection of these to choose from as they are quite popular with professional welders. A standard 6- to 8-inch pair of locking pliers has hundreds of uses around the welding and home workshop.

The next step in the brazing process is, of course, joining the pieces with heat and filling the space between the pieces with molten nonferrous metal. Earlier in this chapter I mentioned that careful surface preparation was about 75 percent of the task of brazing. The remaining 25 percent involves technique, which is what we will cover next.

FIGURE 8.9

Paste flux is brushed on a joint to be brazed.

Brazing Technique

Since brazing is decidedly an oxyacetylene process, the directions for this method of joining metal will be discussed in oxyacetylene terms. As with all oxyacetylene metal joining, it is important when brazing to choose the correct tip size for the project at hand. The type and thickness of metal being worked on will determine the correct tip size to use. Since some metals like aluminum can be heated quite easily, a small tip should be used. It is important to keep in mind that the intention in brazing is not to melt the parent metal but to simply get the joint edges hot enough to melt the filler material and make it flow. You should also keep in mind that a neutral flame has about the same temperature with all size tips, but you may be able to have more control with a small tip. Probably the best place to look for the proper size tip for brazing is in the handbook that came with your torch. Here the manufacturer will have offered suggestions for tip sizes based on the type and thicknesses of the various metals you may come in contact with (see Table 8-1).

Light the torch and develop a neutral flame, which is about best for brazing. Some authorities suggest that a slightly oxidizing flame is better for brazing. In time you may feel the same. I suggest that, when you are relatively new to working with a torch, you work with a specific flame until you have mastered the basics of flame and torch control (Fig. 8.10).

**Table 8-1   Welding tip sizes
and pressure settings for brazing.**

Metal Thickness (inches)	Tip Size (inches)	Size of Brazing Rod (inches)	Oxygen Pressure (psi)	Acetylene Pressure (psi)
1/32	1	1/16	5	5
3/64	2	1/16	5	5
1/16	3	1/16	5	5
3/32	4	3/32	5	5
1/8	5	3/32	5	5
3/16	6	3/32	6	6
1/4	7	1/8	7	7
5/16	8	5/32	8	8

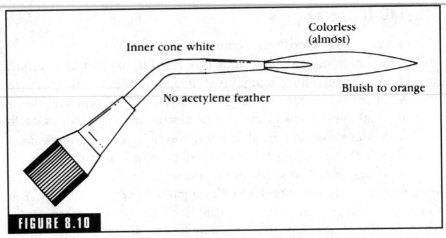

Inner cone white

Colorless (almost)

No acetylene feather

Bluish to orange

FIGURE 8.10

A neutral flame.

Heating the Base Metal

The key to successful brazing is heating the base metal up to a temperature that will be hot enough to melt the nonferrous filler metal but not so hot as to cause the parent metal to become liquid. In most cases this is not as difficult as it may sound, as the melting point of all copper alloy brazing rods, brass, and bronze is between 1,100 and 1,800°F. The melting point of most steels and iron is at least 2,100°F.

It is important that the joint be evenly heated. To accomplish this, it is almost always best to keep the torch moving in a circular, half-moon or zigzag motion over the area being heated (Fig. 8.11). If the joint is a long one, heat up only a section of it at one time. Then proceed along the joint after the initial section has been brazed.

As you heat the metal, watch the flame closely—through welding goggles, of course. You will want to keep the cone of the neutral flame close to the edges of the joint but not touching the metal. Since the tip of the cone is the hottest part of the flame, it may melt the metal if it comes too close to the joint. Overheating is a common problem when learning how to braze (Fig. 8.12).

As the metal heats up, the flux around the joint will give you good indications of the approximate temperature. This is another good reason for choosing the proper flux for the metals being brazed. When heated properly, the flux will boil at about 212°F, leaving a whitish powder on the metal surface. This powder will become puffy and begin to bubble at around 600°F. It will become liquid on the surface

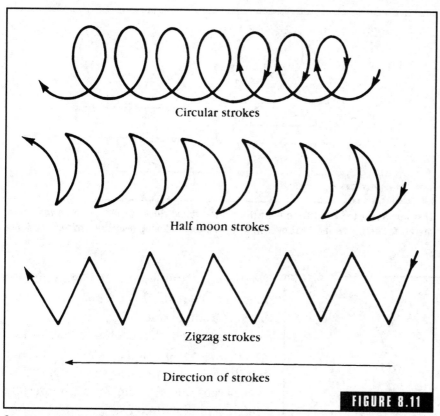

Circular strokes

Half moon strokes

Zigzag strokes

Direction of strokes

FIGURE 8.11

Common patterns of torch movement for brazing.

at about 800°F. At a temperature of about 1,100°F, the flux will once again become liquid and darkish in color. This indicates that just a bit more heat is required to melt the filler rod. The metal of the joint, at this point, will have a dull red color, which is another indication that the temperature is about right to melt and flow the filler rod.

Bonding

When the metal is approximately the right temperature, move the flame away from the joint and touch the end of the filler rod into the joint. If the metal is hot enough, the rod will flow over the heated area and into the joint (Fig. 8.13).

Once the filler metal flows over the heated area, bonding will take place. You

QUICK»»TIP

It is important to keep in mind that both pieces of metal must be of equal temperature because the filler rod will flow only to the hotter piece. This is the main reason for moving the flame over the joint in a circular (or other suitable) motion.

Torch tip

Cone

FIGURE 8.12

It is important to keep the cone of the neutral flame close but not touching the parent metal. To distribute the heat evenly, keep the torch moving over the parts of the joint.

FIGURE 8.13

Apply the brazing rod only after the metal surfaces have been heated up to the point at which they will melt the rod.

should not disturb the brazed joint until the metal has cooled. This will ensure that the bond is a sound one.

After the joint has cooled, you will want to clean up the braze. The amount of cleaning that will be necessary will depend on the parent metals, the filler nonferrous metal, and the flux used. In most cases, only a slight slag or discoloration will be present on the brazed joint. The discoloration can be easily removed with special cleaners or a wire brush. If the slag buildup is heavy, you may have to resort to a chipping hammer and possibly a cold chisel to clean the slag off the joint (Figs. 8.14 through 8.16).

Brazing Versus Fusion Welding

The obvious main advantage of brazing over fusion welding is that it is the best way to join dissimilar metals. Just about any two metals can be joined by brazing except aluminum and magnesium. Another strong point in favor of brazing is that, since the metals need not be brought up to the melting temperature of the parent metals, there is almost never any distortion of the metals. This holds true when brazing is correctly done on any thickness of metal.

Silver and Bronze Brazing

In most instances you will be brazing with a bronze filler rod. One alternative to this nonferrous metal is silver rod. Silver brazing differs only slightly from bronze brazing in that the melting temperature of the silver is lower than it is for bronze. This will mean that less heat is required for the joining and the brazing can be accomplished a bit quicker. Silver rod, as you might expect, is more expensive than bronze. The technique for silver brazing is essentially the same as for bronze brazing.

Brazing Aluminum

Aluminum can also be brazed in the same manner as silver and bronze. Special aluminum brazing filler material and flux must be used for this, however, as brazing aluminum takes place at about the lowest temperatures of all brazing—815 to 1,220°F. Brazing aluminum requires the same general surface preparation as is required when cleaning all other types of metals prior to brazing. Either mechanical or chemical cleaning should be completed before attempting to braze aluminum. Many experts recommend chemical cleaning for aluminum because it is more thorough. Chemical cleaning consists of a series of steps that includes dipping the pieces into a caustic bath, rinsing them, dipping them once again into an acid bath, and then rinsing them. This chemical cleaning is repeated until the surface is totally clean. As with the acid pickling solution mentioned earlier, the

FIGURE 8.14

The work to be brazed must be held securely until after it has been brazed and the joint cooled.

FIGURE 8.15

The bronze rod is applied after the metal has been heated.

FIGURE 8.16

After the brazed joint has cooled, the flux is removed with a wire brush.

QUICK »TIP

To braze using silver, follow this procedure:
- Thoroughly clean the joint surfaces.
- Apply suitable flux to the joint area. Make sure the flux is applied only where you want the silver to flow. Silver rod will not adhere to any surface that is not flux coated.
- Fit the pieces together, and hold them in place.
- Heat the joint area evenly.
- Apply the appropriate brazing alloy (in this case silver rod).
- Let the braze cool undisturbed.
- Clean the surface.

---

ingredients for chemically cleaning aluminum should be available from your local welding supply house.

Since brazing aluminum is done at relatively low temperatures, a small tip size is often the best to use. This will give you more control but not necessarily less heat. It is therefore very important to keep the flame constantly moving over the area being heated.

Braze Welding Technique

The two major differences between brazing and braze welding are the joints between the pieces and the fact that braze welding does not utilize capillary action. Joint design for braze welding is similar to those joints used in fusion welding. Although the pieces must fit together snugly, the close tolerances necessary for successful brazing are not necessary when joining metals with the braze welding process. In effect, the filler metal, which is a nonferrous material as in brazing, is used to fill in the gaps between the pieces. When the filler metal comes in contact with the hot parent metal, it tries to open up the grain. As the filler seeps into the grain of the parent metal, a strong bond is achieved. To be sure, the pieces must fit together well, but not necessarily to the close tolerances required in brazing.

Beveling Metal Edges

Braze welding is a popular means of joining many types of metal using the butt joint. When joining metal that is thicker than ¼ inch, it is necessary to bevel the edges of the joint (Fig 8.17). For metal of lesser thicknesses, it is not necessary to bevel the edges. They should be chipped or ground slightly until bright metal shows. This will aid the filler in adhering well.

Flame Cleaning

All parts should be clean of paint, grease, oil, or any other foreign matter. Solvents and a wire brushing will usually do the job. Another method used to clean metal of surface dirt is called *flame cleaning* (Fig. 8.20). This is simply a process where a neutral flame is passed quickly over the surface of the metal, and it burns off the dirt and other foreign material. This is not a recommended cleaning procedure for metal that has a coating of oil or grease, however, as these will burn violently in the presence of pure oxygen.

After the pieces to be joined have been cleaned, they must be aligned and held in this position during the braze welding pro-

FIGURE 8.17

Metal ¼ inch thick and thicker must have beveled joint edges when braze welding.

QUICK»TIP

Probably the easiest and quickest means of beveling metal edges is on a grinding wheel (Fig. 8.18). You must work carefully, however, as grinding wheels remove metal quickly. It will be helpful if you mark the pieces and the angle of the bevel with a crayon or other suitable marking tool first (Fig. 8.19).

cedure. This can be accomplished in the same manner as when holding pieces in a position prior to brazing, with clamps or locking pliers (Fig. 8.21). On large projects, the pieces are often tack welded at this point, assuming that this welding (in combination with brazing) is not undesirable (Fig. 8.22).

Preheating and Heating

Preheating is often necessary in braze welding to reduce the possibility of expansion and contraction stresses that may be caused by the heat of the braze welding operation. Preheating can be accomplished with a neutral flame. When the metal turns a dark red color, it is about right. Keep in mind that preheating is usually only necessary with the thicker pieces of metal, say, over ¼ inch.

It is important when braze welding to keep the torch flame moving over the surface of the joints. This will ensure that the metal is heated evenly along the joint. The flame, as mentioned earlier, is neutral, and

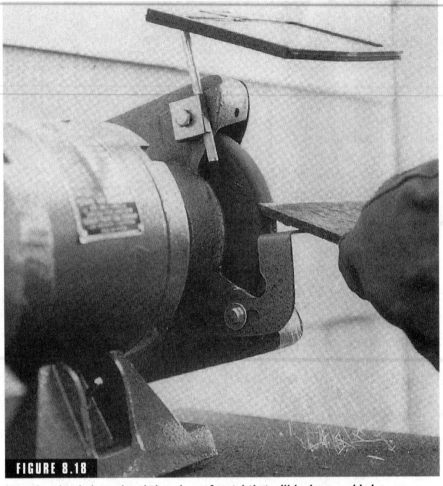

FIGURE 8.18

Use a bench grinder to bevel the edges of metal that will be braze welded.

the cone of this flame should not be allowed to touch the metal because the parent metal may melt.

While heating the joint surface, the end of the bronze welding rod should also be heated and then dipped into the powdered flux. When the bronze rod is hot enough, the flux will adhere to and coat the rod.

Tinning

When the parent metal is very dark red, it is about time to push the end of the bronze rod into the joint. This will allow some of the flux to melt onto the surfaces. As the parent metal melts the bronze rod, it will form a thin coat on the metal surfaces (Fig. 8.23). This is known as *tinning,* and it is similar to the tinning process used in soldering. When tinning the surfaces of the joint in braze welding, you will find a rubbing action the best way to get the flux and bronze to stick to the parent metal.

FIGURE 8.19

Mark metal with a soapstone pencil before grinding. Use the mark as a guide to the proper bevel angle.

You must be careful not to heat the parent metal too much as this will cause the filler metal to bubble. Good tinning will then be impossible. On the other hand, if the metal is not heated enough, the filler metal will form little balls on the surface and not in the surface.

Once the surfaces of the joint have been completely tinned, you can go back over the joint and add more filler metal in one or more passes. It is very important that you work continuously, tinning and then building up the filler metal. If the filler metal is allowed to cool, it may break down when reheated. The end result will be a very poor quality braze weld (Fig. 8.24).

FIGURE 8.20

Flame cleaning metal.

Cooling

Even cooling is important when braze welding. There are several ways of accomplishing this, including heating the parent metal around the joint at least twice as wide as the distance of the joint. For example, if the width of the joint is ½ inch, heat the parent metal on both sides of the joint for a distance of about 1 to 1½ inches away from the joint. Metal heated in this way will cool slower than if just the joint area is heated.

Another way to slow down and make the cooling more uniform is to cover the welded piece with asbestos cloth. This will cause the piece to cool very slowly.

Still another way of cooling, and a method often used in industry when working with cast iron, is to pack the hot project in ashes or

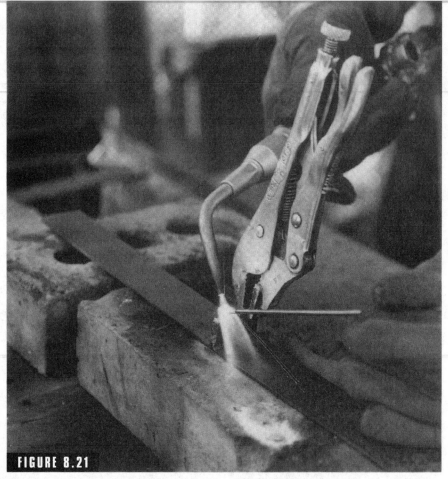

FIGURE 8.21

Vise-Grips are handy for holding a joint steady while brazing.

lime. This will almost guarantee that the metal cools very slowly, thus creating more ductility in the welded material.

After the metal has cooled, it may be necessary to clean up the welded area. This is done to remove any flux residue or slag. A wire brushing is usually all that is required for this cleaning.

General Comments

Brazing and braze welding techniques are relatively easy to learn and master, providing that the do-it-yourself welder devotes the necessary time to surface preparation and flame control. Undoubtedly the best way to learn how to braze and braze weld is to practice on scrap pieces of steel until you have a clear picture of what is necessary for success. Since brazing is, in effect, an extension of soft soldering, this may be a good place to learn the basics. Begin by learning how to make simple lap joints using a soldering iron, flux, and solder. Work on sheet metal until you have an insight into the subtleties of heating metal. Once you understand this method of metal joining, move on to thicker metal, and use an oxyacetylene torch to braze the pieces together.

In the beginning, it may be easier to work on the thinner gauges of steel, say, 1/8 inch thickness or less. Here again, you should practice the lap joint as it is probably the easiest to master.

After you have mastered the light touch necessary for successful brazing and the lap, butt, and scarf joints, move on to braze welding. You must remember that part of the key to braze welding is applying a thin coat of the nonferrous filler material to the surfaces of the joint, known as *tinning the surface.* It is very important that this tinning coat

FIGURE 8.22

Tack welding the edges of a joint.

FIGURE 8.23

A thin coat of bronze on a metal surface.

PROCEDURE CHECKLIST

___ Clean the surfaces to be joined.

___ Fit the pieces together closely, and use the proper flux.

___ Heat the metal evenly by constantly moving the neutral flame over the joint area.

___ Keep an eye on the flux as this will be one of your best guides to the temperature of the metal.

___ When the metal is hot enough, remove the flame and introduce the brazing rod. It should flow into the joint because of the temperature of the metal, not from the heat of the flame.

___ Let the brazed joint rest undisturbed until cool.

___ Clean the cooled joint with a wire brush to remove flux and slag, if necessary.

be soundly attached to the surfaces of the joint metal. This requires that the surfaces be brought up to the proper temperature before applying the filler rod.

In the beginning your brazed joints may not stand up under strain. As you become more familiar with the various metals and temperature ranges, you should be able to produce professional-looking finished joints (Fig. 8.25). Once you have mastered the techniques of brazing and braze welding, the obvious next step is to learn how to weld.

FIGURE 8.24

Joining thicker metal with two-pass brazing.

FIGURE 8.25

Comparison of good and bad joints for brazing.

Beginning Welding

Up until this point, we have discussed two methods used for joining metal: soldering, which takes place at temperatures less than 800°F, and brazing, which requires flame temperatures from 800°F to just below the melting point of the metal being joined. In both cases, the metals being joined are not brought up to a temperature that exceeds the melting point of that particular metal. Now we will deal with the final method of joining metals and one that produces the strongest attachment: welding.

Fusion Welding

We use the term *welding* rather loosely here to describe the process whereby two pieces of similar metal are heated up to their particular melting point. The molten metal then flows together forming one homogeneous piece. In actuality, this process is called *fusion welding*.

A sound welded joint is a thorough mixture of the base metal and, in many cases, a filler rod made from a similar material. Regardless of the thickness of the metal, a good welded joint will be fused through-out the joint. When joining relatively thin pieces of metal, less than $1/4$ inch thick, making a thoroughly welded joint is not difficult. But when joining thicker metals, special joint preparation techniques must be employed. As with both soldering and brazing, joint preparation is a necessary part of fusion welding.

Obviously, the best way to learn how to weld is to practice often and on various thicknesses and types of metals. It must be assumed, at this point, that you are familiar with your oxyacetylene equipment. You should know, for example, how to set up the equipment and develop the various types of oxyacetylene flames. These are the acetylene flame, carburizing flame, neutral flame, and oxidizing flame (Fig. 9.1). You should also have developed your own technique for keeping the torch flame in motion and at a specific height above the work. These techniques were discussed in Chapter 8, but a quick review here may be of help at this point.

Torch Movement

In order to ensure that the metal or joint between the two pieces of metal is heated evenly, the flame of the torch must be kept moving so that the metal will be brought up to a uniform temperature and will flow at the same time. This is best accomplished by moving the torch tip continuously. Figure 9.2 shows several methods or patterns, if you will, of moving the torch to distribute the heat evenly. At this time, at least one of these torch movement techniques should be familiar to you. More often than not, I find a circular movement the most effective and easiest to perform.

For welding metal, there is another technique that you will have to perfect. This technique also concerns torch movement, either forward

		Neutral	Oxidizing	Carburizing	Aircowelding
Torch flames		Luminous cone —5850°F Envelope —3800°F —2300°F	—6300°F	—5700°F	—5800°F
Ratio	Oxygen acetylene	$\dfrac{1.04-1.14}{1}$	$\dfrac{1.15-1.70}{1}$	$\dfrac{0.85-0.95}{1}$	$\dfrac{0.92-0.98}{1}$
Effect on metal		Metal is clean and clear, flowing easily	Excessive foaming and sparking of metal	Metal boils and is not clear	Similar to neutral flame—little or no puddling necessary

FIGURE 9.1

Characteristics of oxyacetylene welding flames.

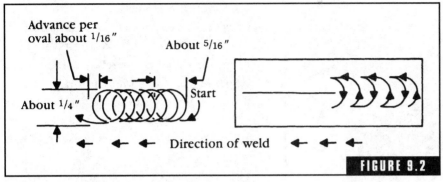

Advance per
oval about $1/16''$

About $5/16''$

Start

About $1/4''$

← ← ← Direction of weld ← ← ←

FIGURE 9.2

Two common blowpipe movement patterns.

or backward, along the joint. Actually there are two methods for general torch movement: forehand welding and backhand welding.

When using the *forehand welding method,* the torch flame moves along the joint toward the tip of the welding rod. The position of the welding rod is actually on the cooler area, not yet molten area, of the joint. The flame of the torch is pointed in the direction of the welding but also slightly downward so the flame preheats the joint area (Fig. 9.3). The forehand method works well with thin, sheet type metal.

Direction of weld

Forehand welding

FIGURE 9.3

Forehand welding technique.

The *backhand welding method* is opposite from the forehand welding method. Actually the flame of the torch precedes the welding rod in the direction the weld is being made. When backhand welding, the welding rod is used to work the puddle of molten metal. As you can see in Fig. 9.4, this differs significantly from the forehand method of welding.

Generally speaking, the backhand method of welding offers greater control of the welding process while at the same time providing welds with equal strength. In most cases it is the preferred technique for welding metal that is thicker than about $1/4$ inch.

Direction of weld

Backhand welding

FIGURE 9.4

Backhand welding technique.

QUICK>>>TIP **Throughout this chapter, unless otherwise stated, the flame being used for the welding process is a neutral flame. You may recall that the neutral flame is the third type that normally develops when a torch is fired up (Fig. 9.5).**

As I mentioned earlier, the best way to learn how to weld well is to weld often. In the beginning of your welding practice, work on thin pieces of metal, $1/8$ inch or so thick. After you have mastered the various techniques necessary to weld metals of this thickness, move up to thicker pieces. Welding thicker metals is discussed in Chapter 10.

One very good way to learn some of the basics of heating metal up to the melting point is to practice on a single piece of sheet steel. In the beginning, work with steel that is no more than $1/8$ inch thick. This metal is easily heated and is quite suitable for learning the techniques of welding.

Puddling Metal

As your first exercise, begin by placing a 6- to 8-inch-long piece of steel between two firebricks. Set up your welding equipment, and attach a small welding tip to the blowtorch handle. Consult the guide that came with your unit for the proper oxygen and acety-

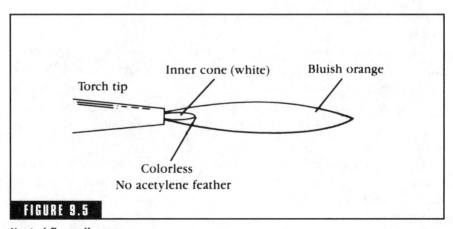

Inner cone (white)   Bluish orange

Torch tip

Colorless
No acetylene feather

FIGURE 9.5

Neutral flame diagram.

lene settings for the tip being used. Light the torch, and develop a neutral flame. It is assumed, at this point, that you have put on suitable clothing, gloves, and tinted welding goggles.

Heating Steel

Begin heating the steel in about the middle of the piece (Fig. 9.6). You should hold the torch in a comfortable and balanced position. The tip should be slanted at about a 45-degree angle to the right (southpaws should make suitable adjustments). The direction of travel will be in a circular motion from right to left. The circles you create should be about ¼ inch in diameter. The tip of the cone of the flame should be ¹⁄₁₆ to ⅛ inch above the metal.

Watch the metal closely as it begins to heat up. It will turn cherry red, orange, yellow, and finally white. About the time the metal turns white hot, it is in a molten state. While heating the metal up to this state, it is important that you keep the flame cone moving over an area of about ¼ inch. If you simply keep the flame in one spot, that spot will turn white and finally become a hole (Fig. 9.7). Since this is only practice, you should experiment with both holding the torch in a fixed position, which will burn a hole in the metal, and moving the torch in tiny circles causing the molten metal to puddle.

When the metal is in the molten state, you can move it around slowly with the cone of the flame. As you move the torch tip, you will notice

FIGURE 9.6

Begin heating the steel in the middle, and watch the metal change color as it gets hotter.

FIGURE 9.7

Hold the flame in one spot until you have burned a hole in the steel.

that the molten puddle can also be moved. This, in fact, is one of the basic welding procedures. Much of your later welding will be carried out in a similar manner. In some cases, you will be using a filler rod, while in other instances you will simply be welding with the existing metal, as we have just done in this puddling exercise.

Puddle Characteristics

There are a number of things that can be learned by carefully watching the puddle of molten metal. For example, the diameter of the puddle will be in proportion to its depth. On thin metal the puddle will or can be larger than on thicker metal. In the latter case, you must keep in mind that the heat necessary to create molten metal has to penetrate through the thicker metal, which is a large part of the reason for the smaller puddle. Thicker metal also requires a larger tip size to help create more heat.

The general characteristics of the puddle will give you a very good indication of the type of flame you are working with. A neutral flame will produce a puddle that appears smooth and glossy (assuming, of course, that the grade of metal is a good one, with few impurities or dirt). If the puddle sparks or bubbles excessively, this is usually an indication that the metal is dirty or of poor quality. The puddle will appear dull and dirty if the flame is more carburizing than neutral.

When a properly adjusted neutral flame is being used to create a puddle of molten metal, you will notice a small bright incandescent spot on the edge of the puddle opposite the flame. This spot will move quite actively around the far edge of the puddle. If the flame is not neutral, the spot will be oversized and will barely move.

Once you have learned how to develop a puddle of molten metal and have burned a few holes in this scrap piece of metal, the next step is to move the puddle over the surface of the metal. A brief description of the techniques used will be of some help.

Welding Beads

The torch tip, with a neutral flame, is ¹⁄₁₆ to ¹⁄₈ inch above the surface of the metal and held at about a 45-degree angle to the surface. The direction of travel will be from right to left and in a circular motion (Fig. 9.8).

Begin forming a puddle just in from one edge of the sheet steel. As the metal begins to become molten, slowly advance toward the far end of the piece in a circular motion. As you move forward, the metal will go from a solid to a molten state. Then, as you move toward the opposite edge, the steel will become solid once again. You will notice that the circular movement of the torch will produce circular ridges as the metal cools. This line of ridges is commonly called a *welding bead.* Develop this welding bead across the face of the steel (Fig. 9.9).

FIGURE 9.8

After you have burned a hole in the steel, try heating the metal once again in another spot. Try to puddle the metal in a straight line.

With a little practice you should be able to create a welding bead that is uniform and straight. In the beginning, you will undoubtedly burn holes in the metal and probably develop a bead that more than slightly resembles a snake. After a bit of practice, however, you should be able to create a welding bead that is both uniform and straight.

FIGURE 9.9

A finished weld done without a welding rod.

QUICK»TIP   As a side note, it makes a lot of sense to keep a pair of heavy-duty pliers handy while welding. The pliers can be used for picking up the hot metal. This will enable you to inspect the weld safely before the metal has cooled to the touch (Figs. 9.10 and 9.11).

Penetration, or depth, of a weld is very important. It is entirely possible to form a puddle of molten metal on the surface and not all the way through. When practicing, you should turn the scrap piece of steel over to see that the weld has penetrated through the metal.

You should practice puddling sheet steel until you can create a straight, well-penetrated welded bead across the face of the steel. Most knowledgeable welders will probably agree that if you can create five straight and well-formed welding beads, one after another, you are probably proficient enough to try more serious welding (Fig. 9.12).

Working with Welding and Filler Rods

Until this point we have been practicing only with a single piece of steel and a flame. Since much welding uses a welding or filler rod as well, this will be our next objective. Since working with a welding torch and filler rod requires that you use both hands, the following exercise should not be attempted until you have reasonably mastered working with a torch to create a welding bead by puddling.

At this point, some mention should be made about welding or filler rods used by the oxyacetylene welder. Often during the oxyacetylene welding process, more metal than that which is melted by the torch is needed for a successful weld. To add more metal, it is common practice for the welder to use a welding rod that has the same, or very similar, properties as the base metal that is being joined.

Welding rods are available in many different alloys and designed for just about any type of joining task. Probably the most common type of welding rod is simply mild steel. These are commonly sold in 36-inch lengths and are copper coated to protect the rod from rust and corrosion while in storage. Uncoated, mild-carbon steel rods are also available. If there is an all-purpose welding rod, a mild-carbon steel rod is probably the one to buy. Diameters range from $1/16$ to $1/4$ inch. Your local welding supply house will have a selection of welding rods. You should buy an assortment of these mild-carbon steel, copper-coated rods. In most cases, you will use

FIGURE 9.10

Pick up the steel with a pair of pliers and examine the cooled puddle.

FIGURE 9.11

Be sure to check the back side of the steel to see how well the puddle penetrated through.

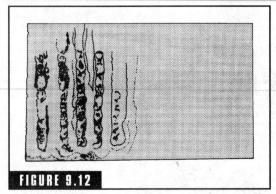

FIGURE 9.12

Practice puddling the steel with the torch until you can produce five straight and uniform puddles with good penetration.

FIGURE 9.13

An assortment of welding rods.

the 1/8-inch-diameter rods with the greatest frequency (Fig. 9.13).

Coat Hangers

Some mention should be made about the economy of using a substitute to the all-purpose welding rod. If you spend enough time in the company of welders, you will probably hear that a standard wire coat hanger can be used as a substitute for mild-carbon steel welding rod. While it may be true that a common coat hanger can be used as a filler material when welding, you should keep in mind that the steel used for these hangers is generally of a very low quality. It is therefore not wise to expect that a wire coat hanger will hold up under stress. An example may help to illustrate the value of using a standard welding rod instead of a wire coat hanger.

An acquaintance was restoring an antique automobile. One of the many tasks that needed to be accomplished was the installation of new motor mounts in the engine compartment. It was about this time that the friend decided to purchase a new welding outfit and use it to help him in his restoration. After becoming reasonably competent at welding, he decided that it was time to start using the welding equipment for attaching new motor mounts. I must admit that he did an impressive job of welding the mounts onto the frame. The finished welds looked as if an old pro had done the work. The only problem with the project was that the welded mounts did not hold up under the torque of the engine because he used wire coat hangers instead of an all-purpose welding rod as a filler material. To make a long and rather expensive story short, my friend had to reweld the mounts. The second time around, however, he used standard copper-coated steel welding rods, and the welds have held to this day.

My friend is all that much wiser when it comes to welding. You can bet that he will never use wire coat hanger as a filler material ever again. In fact, in his entire household, you cannot find any type of clothes hanger other than a wooden one.

Adding Filler Material

To produce a welding bead with the aid of a welding or filler rod, you begin in very much the same manner as when working solely with a torch. The major difference, of course, is that in addition to holding the torch in your right hand, you will also be holding (and moving) a welding rod in your left hand. The welding rod is used to add material to the puddle of molten metal. The more proficient you can become at adding filler material to a weld, the better your overall welding and the stronger your welds will be.

FIGURE 9.14

Practice holding the torch and welding rod in the proper position before lighting the torch.

It is important that you become proficient at adding filler rod to a puddle of molten metal. To practice, begin in the same manner as the first exercise in this chapter, with a single piece of sheet steel suspended between two firebricks. Hold the torch in your right hand and a welding rod in your left (Fig. 9.14). Begin heating the metal with small circular motions in about the middle of the piece of steel. As the metal begins to pass through the different color changes, slowly lower the tip of the welding rod into the flame to preheat it (Fig. 9.15). It is important that you do not touch the cone of the flame with the welding rod.

As the parent metal begins to become molten, the tip of the preheated welding rod is touched to the edge of the puddle. Almost immediately, the welded rod will melt into the puddle and mix with the parent metal (Fig. 9.16). After enough welding rod has been added to the puddle to create a slight

FIGURE 9.15

As you heat up the metal, keep the tip of the welding rod in the outer flame to protect it.

FIGURE 9.16

As the parent metal becomes molten, touch the pre-heated tip of the filler rod to the far edge of the puddle. The rod will then flow onto the surface of the metal.

FIGURE 9.17

A good weld bead produced with a filler rod.

crown on the surface, the tip of the welding rod should be withdrawn to about ⅛ inch above the surface of the welding bead.

Torch control is very important when introducing filler material to a weld. The flame should be in constant motion as should the tip of the welding rod. The cone of the flame does not cover a large area, but it is moving nevertheless. This will ensure even heat over a small area.

Practice melting filler rod into a weld bead until you can produce five straight welding beads. Probably the most difficult technique to master when introducing filler rod into a welding bead is the circular motion of both hands. This technique can be mastered only through practice. It is extremely important that you practice adding filler material to the molten surface of the parent metal until you can produce a respectable welding bead (Fig. 9.17).

Torch Control

Torch control when adding filler material is also quite important for a satisfactory weld. You should remember that the tip of the torch should be held at about a 45-degree angle to the

TRADE SECRET At this time a few professional pointers are in order. It is important that the tip of the welding rod be pre-heated before it is touched to the surface of the puddle of molten metal. A rod that is outside the flame envelope will be much cooler than permissible (Fig. 9.18). When it is touched to the puddle, it will cause the molten metal to cool. The end result will be a poor-quality weld bead.

On the other hand, if the tip of the filler rod is held too close to the cone of the flame, it will melt and drip into the puddle. More often than not, this will result in the filler material being blown away from the weld, which will produce an uneven bead and, more importantly, incomplete fusion.

FIGURE 9.18

It is important to keep the tip of the filler rod in the outside of the flame envelope.

surface of the parent metal (Fig. 9.19). When the torch is being held correctly with the proper flame, the surface of the puddle will be glossy and smooth. Often a slight angle adjustment is all that is required to produce a puddle with these characteristics.

Filler Rod Size

Another consideration for a successful and sound welding bead is the size of the filler rod. Table 9-1 offers some suggestions as to the diameter of the rod to use for the various thicknesses of metal. In the beginning, use the recommended welding rods for the particular job at hand. As you work with the different thicknesses of metal, you will develop a familiarity with how a particular diameter rod performs. You will therefore be able to create a suitable welding bead.

Some points to consider concerning improper diameter welding rods are worth mentioning at this time. If a $3/32$-inch welding rod is the proper one to use for a particular joint and you switch to a thinner rod, say, $1/16$ inch, you will find it very difficult to add enough filler material to the weld bead. A smaller rod will also make controlling the puddle and resultant bead very difficult as well. The smaller rod will also tend to oxidize or burn very quickly.

Using a rod that is larger than what is called for can have similar effects on the finished weld. For example, if a heavy rod is used on

FIGURE 9.19

The torch tip should be held at a 45-degree angle, and the flame should be about $1/8$ inch off the surface of the parent metal.

Table 9-1 Welding Rod Sizes

Metal Thickness (inches)	Welding Rod Diameter (inches)	Oxygen Pressure (psi)	Acetylene Pressure (psi)
$1/16$	$1/16 - 3/32$	4	4
$1/8$	$3/32 - 1/8$	5	5
$1/4$	$5/32 - 3/16$	8	8
$3/8$	$3/16 - 1/4$	9	9

light metal, it may create too large a puddle and therefore a sloppy weld bead. A filler rod heavier than required may also tend to cool the puddle of molten parent metal too much while it is being added. The end result will be poor penetration and an ineffective weld bead.

Joining Two Pieces of Steel

Once you can create a straight weld bead on the surface of a single piece of sheet steel, it is time to begin practicing joining two pieces of steel. For this exercise select two pieces of ⅛-inch-thick steel strap, about 8 inches long. Lay the two pieces, with long edges aligned and butted together, between two firebricks. In order to hold the two pieces together while welding, the ends should be tack welded. *Tack welding,* as the name implies, simply means to spot weld one very small section of the pieces being joined. In addition to being a good way to learn how to heat up two pieces of metal to the melting point, tack welding is also a very useful means of holding two pieces of metal together while they are being welded along their length.

Tack Welding

Hold the lighted torch so that it balances in your right hand. Move the flame close to the end of the two strips—that area where the two strips butt together. Move the cone of the flame slightly in a tiny circular motion over the joint so as to cover an area of about ⅛ to ¼ inch. Watch the metal change from its normal color to red and finally to white hot.

Just about the time that the metal turns white in color, it should begin to flow together. As soon as this happens, you must move the flame away from the joint. It is important to keep in mind that if the heat continues to be applied to white-hot and flowing metal, you will often cause a hole to appear. This is undesirable. Your aim is to simply bring the edges of the two pieces of metal up to the melting point, and then just a bit more so the edges fuse together and become one. Circular torch movement usually helps in fusing the pieces (Fig. 9.20).

Once a tack weld has been completed on one end of the strips, move the flame along the joint between the pieces and back a few times to heat up the metal slightly. Then tack weld the opposite end of the two pieces. Make sure that the pieces remain aligned while doing this tack welding.

In case you are wondering why the metal is heated up after one tack weld has been completed, this is so that the two pieces will cool at approximately the same, even rate. This even cooling will reduce the chances of the strips buckling or warping as they cool.

Prefitting Pieces

There is nothing difficult about tack welding. But there are a number of procedures used by professional welders to ensure the success of the weld. To increase your chances of producing a sound tack weld, you should know that the pieces to be tacked must fit tightly together

FIGURE 9.20

Tack weld the joint edges of the two pieces without a filler rod.

along their entire length. Prefitting will make heating the two edges easier because you will not have to move the torch flame over a very large area. It will also enable you to flow the two metals a bit easier. If the two edges are not aligned snugly, you can still produce a strong tack weld. But you will find it necessary to use a bit of filler rod to make up the difference between the pieces.

Welding Irregular Edge Pieces

In some cases, it may be necessary to make a tack weld with the addition of a filler rod. One example is when the two pieces being joined have irregular edges. To accomplish this, begin by aligning the edges of the two pieces as closely as possible. Working on one end of the piece, heat both edges up to the white-hot stage and add the filler rod. As you work, you will find it necessary to keep the flame moving constantly. Add the filler material at the exact moment the metal begins to flow. If you wait too long, the metal from either or both pieces may begin to drip, necessitating adding even more filler rod.

After the second tack weld has been completed, the strips will remain in position while you weld the long joint between the two pieces. Your first attempt at welding two pieces of metal together should be done without the aid of any filler rod. For the record, when tack welding long pieces of metal more than 12 inches in length, the edges are tack welded along the entire length, about 6 inches apart.

Heat Application

Begin at one end of the joint between the strips. With small circular movements of the torch, heat up the edges of the joint until they are in a molten state and are fusing together. As the joint begins to fuse, slowly move the flame toward the far end of the joint to ensure that complete fusion is taking place. It is important to keep the cone of the flame moving at all times, while at the same time working toward the far end of the joint. When working with thin metal, such as $\frac{1}{8}$ inch thick or less, you will find that you can move relatively quickly along the joint (Fig. 9.21). With heavier metal, you will find it necessary to work more slowly and with a larger size tip.

In principle, this exercise in welding is the same as the first welding exercise in this chapter on puddling metal. The major difference, in this case, is that you are puddling the edges of two pieces of metal rather than simply working on a single piece of steel.

As you progress along the joint, it is important that you apply enough heat to fuse the edges of the pieces all the way through. Your

FIGURE 9.21

Puddle the edges of the joint without the aid of a filler rod. Keep the torch flame moving in tiny circles as the metal becomes molten.

FIGURE 9.22

After you have made one puddling pass along the joint without a filler rod, turn the joined pieces over with a pair of pliers, and check the penetration of the weld.

best indication of the penetration of the weld will be the puddle being created by the cone of the flame. It will be helpful to you, especially at this early stage of metal joining, to stop welding and turn the pieces over to make certain that the weld has fused the edges of the pieces all the way through. Use a pair of pliers or a gloved hand for turning the metal over. If a circular weld appears on the back side of the joint, you are applying enough heat. However, if the weld does not show on the other side, you are not applying enough heat to the joint (Fig. 9.22).

Welding Joints

Practice welding the joint between two pieces of light steel until you can accomplish this task with some competence. See if you can produce a strong weld without the aid of a filler rod. After you have mastered welding two pieces of light steel, where the butting edges are on the same plane, you should turn your attention to welding other types of joints, still without the addition of a filler material.

FIGURE 9.23

An outside-corner weld with the vertical member extending $1/16$ to $1/8$ inch above the horizontal member.

Outside-Corner Weld

Another type of welded joint that does not necessarily require the addition of a filler rod is called the *outside-corner* weld. For this exercise select two pieces of light steel, $1/8$ inch thick, 8 inches long, and 2 to 4 inches wide. Place them together at right angles so that the vertical piece extends beyond the horizontal piece by about $1/16$ inch (Fig. 9.23). Begin by tack welding both ends of the strips to hold them in position while you weld the joint. The metal that extends will be used as a filler material as you weld.

A few professional pointers will help you to achieve success when making the outside-corner weld. Begin by holding the torch so that the flame plays on the horizontal piece. This will cause the weld to form on the horizontal surface where it belongs. Cleaning up the weld bead will be much easier later on as you will only have to grind one side, the horizontal piece.

When making an outside-corner joint weld, you will find that less torch movement is necessary than usual, particularly on butt joints (Fig. 9.24). You still must move the cone of the flame so that it heats up both pieces, but your main intention will be to "flow" the extended vertical piece onto the horizontal piece. This generally means less torch movement. You will also find that working from the outside of the vertical piece is the best way to approach this type of weld.

FIGURE 9.24

Proper torch position when welding an outside-corner joint.

After the outside-corner weld has been completed and the metal has cooled and solidified, check the weld first for appearance and then strength. A careful look over the weld bead will give you a good indication of how sound the weld is. Next, check the strength of the weld by trying to bend the two pieces of metal, in the same way you would open a book (Fig. 9.25). If you can hear any cracking or breaking of the metal, then the weld did not penetrate very deep into the joint between the pieces.

One of the unique features of the outside-corner welded joint is that the weld need not be visible on the inside of the joints. In fact, the weld should be strong enough without penetrating all the way through. The best way to discover if the weld is, in fact, strong enough is to try to bend the two pieces apart. If the pieces hold without giving in the slightest, the outside weld is a good one.

FIGURE 9.25

Test the strength of an outside-corner joint by trying to bend the pieces, in the same way you would open a book.

FIGURE 9.26

A flange joint held by a pair of Vise-Grips.

FIGURE 9.27

A flange joint being welded without filler rod.

Flange Joint Weld

Still another type of weld that can successfully be made without the aid of a filler rod is called a *flange joint weld.* To make this type of joint, you must first bend both edges of the pieces to be welded so that each piece will have an edge with a 90-degree angle. The flange should be from $\frac{1}{4}$ to $\frac{1}{2}$ inch wide and should be roughly equal to the piece it will butt against (Fig. 9.26). You will find a heavy-duty bench vise quite handy for bending the edges of sheet steel.

Begin a flange welded joint by aligning the flanges of the pieces and tack welding both ends to hold the pieces securely while welding. Next, apply heat from the torch to the top of the flanges until they melt and fuse together. As the two flanges melt together, enough filler material from the parent metals should be present to produce a strong and sound weld (Fig. 9.27).

As you can see, there are a number of welds that can be accomplished without the addition of a filler material. Generally speaking, this is the case more often than not when working with thin steel. As the thickness of the steel increases, however, you will find it more difficult to develop a built-up welding bead. In fact, at this time you may have discovered that the bead you can produce by simply puddling the metal is slightly indented. Whenever this happens, it means that additional filler material is required. For our purposes, that will mean welding rod. It will probably be most helpful if I describe how to make some of the more simple welds using a filler material on relatively thin steel. Keep in mind, when adding filler material in the form of welding rod, that we are approaching the joining process in very much the same manner as when joining metal without a filler rod, that is, pud-

dling the metal. The major difference, of course, is that now we are
building up the weld bead with the filler material. Basically, everything else is the same.

Butt Joint Weld

Since the butt joint is the easiest of all joints to accomplish, you should
begin welding with filler rod with this simple joint. Lay the two pieces
of sheet steel, about ⅛ inch thick, side by side and suspended between
two firebricks. Next, tack weld both ends of the strips to hold them in
position while you weld the joint. Begin, if you are right handed, on the
right end of the two pieces and work toward the left. For welding metal
less than ¼ inch thick, use the forehand welding method described earlier in this chapter. To briefly refresh your memory, the forehand
method of welding has the welding rod held in position over the as-yet-
unwelded section of the joint. The flame of the torch is held at about a
45-degree angle to the surface of the joint. The flame is played on the
joint in a circular motion, while the tip of the welding rod is held

inside the flame envelope to preheat it. As
soon as the metal edges of the joint become
molten and begin to flow together, the tip
of the welding rod is touched to the far
edge of the puddle. This will cause some of
the filler rod to flow into the joint and
become part of the weld bead (Fig. 9.28).

You must move smoothly along the
seam, constantly keeping the flame moving to evenly heat both sides of the joint.
In the beginning you may find it difficult
to hold the torch at a constant height
above the joint while at the same time
holding the tip of the welding rod in the
flame envelope. If you have a problem
coordinating both hands, try locking your
elbows against your sides and keep your
forearms at a 90-degree angle to the front
of your body. In time you will develop
this technique. When you are first starting, it seems difficult, hedging on the
impossible (Fig. 9.29).

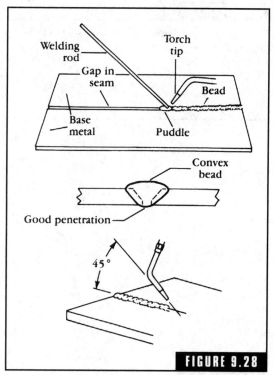

FIGURE 9.28

The filler rod is applied as the weld metal becomes
molten and starts to flow.

FIGURE 9.29

Try locking your elbows against your sides to steady both the rod and blowpipe actions.

FIGURE 9.30

Welding a T joint using the forehand welding method.

As you move along the joint, it is important to keep the flame moving at an equal distance over both edges. You will not be covering a very long span at any one time. Instead, make sure that you are creating a good puddle of molten metal. When the edges of the joint begin to fuse together, lower the tip of the preheated welding rod into the puddle. Allow the filler rod to flow into the puddle until the height of the welding bead is built up slightly. Then proceed a little bit further along the joint and repeat the heating. Your aim, when adding the welding rod, is to build up a joint that is equal along its entire length.

Whenever you come upon a tack weld, heat the area as if it did not exist, and add the filler rod. During the entire weld, the filler rod should be added in uniform amounts and at regular intervals. It is extremely important that you do not try to rush through the welding. Allow the edges of the metal to heat up sufficiently and to flow so good penetration and fusion can take place.

Practice the butt weld, with the addition of filler rod, until you can produce a respectable weld bead with good penetration through the joint. Work only with sheet steel that is less than ¼ inch thick until you have perfected and mastered the torch and rod movements. After you can make a decent butt weld, the next exercise should be to learn how to make a *T-joint weld*.

T-Joint Weld

T joints are formed by placing one sheet of steel on a similar piece of steel so that the two pieces resemble Fig. 9.30. The next step is to tack weld the ends to hold the pieces in position while you weld the joints. After the tack welds have cooled for a few minutes, the next step is to weld the vertical piece to the horizontal piece. The forehand method of welding should be used initially. Once you have reasonably mastered this technique, try the backhand method.

When welding a T joint, it is important that you do not allow too much heat to build up on the vertical piece or you may burn through it. While the flame of the torch should be kept in motion during this welding, you will probably find that the pattern of movement will be very tight. Lower the tip of the welding rod into the joint when the metal of both pieces starts to flow. It is important at this time to keep an eye on the vertical piece as it will always have a tendency to reach a molten state quicker than the horizontal piece. If you find this to be a problem, adjust the angle of the flame so it is closer to vertical than horizontal (Fig. 9.31).

After the joint has been welded uniformly along its length, allow the metal to cool, and then inspect the weld. The bead should be built up to an equal height along its entire length and be equally distributed along both pieces. In other words, one side should not have more bead or filler material than the other. The vertical piece should appear to be solid and the same thickness as when you began. If the vertical piece looks thinner in certain areas, it is usually an indication that you have melted some of the metal away. The thinner area would not have as much strength as the rest of the joint (Fig. 9.32).

FIGURE 9.31

Adjust the torch angle more toward 90 degrees so that you will have better control when making a T joint.

FIGURE 9.32

Clean up the joint with a wire brush.

Welding Thicker Metals

We have covered how to weld relatively thin pieces of steel both without and with the addition of a filler material. In just about every instance, thinner metal does not require any edge preparation other than making sure that the joint fits together well and it is clean. When working with thicker metals, however, you will generally not be able to achieve good weld penetration unless you reduce the thickness of the metal along the joint area. It is generally accepted that when welding any metal that exceeds ¼ inch in thickness, you must reduce the thickness of the joint edges by grinding. Then fill in the area with welding rod. Actually, the point of edge treatment is to ensure that the heat from the oxyacetylene flame will be even along the joint. The thicker the metal along the edges, the more heat must be applied to make the edge metal molten throughout. As it turns out, metal thicker than about ¼ inch cannot be heated thoroughly enough to make fusion welding complete and total. This is the basis for edge treatment.

Basically, there are four types of joints in all welding: butt, T joint or fillet, flange, and lap. Since we have covered these previously, the information will not be duplicated here (Fig. 10.1).

Beveling Metal Edges

When working with thick metal, some type of edge treatment must be initiated before welding begins. In many cases this will simply amount

FIGURE 10.1

Common joint designs for metals less than ¼ inch thick.

to beveling one or more of the edges at about a 45-degree angle. Then, when the two pieces are placed together prior to welding, the beveled edges will form a V, which can then be welded. In some cases the depth of the V groove will be deep, which will require that more than one pass be made with torch and filler rod.

There are really only a few ways in which the edge of a piece of metal can be beveled. These include *grinding, cutting,* and *machining.* In most cases only the first two are within the means and capabilities of the average home welder, machining being a commercial production form of edge preparation.

Grinding

Grinding the edges of metal is probably the easiest and most efficient means of edge beveling for the do-it-yourself welder to perform. Since every home workshop should have some type of grinding wheel, chances are good that you will already possess the equipment necessary for this special edge preparation. Begin by marking the angle of the bevel, in most cases 45 degrees, and press the metal into the spinning grinding wheel. For larger pieces, you may find the work easier to accomplish if you set up some type of guide or rest on the front of the grinding wheel (Fig. 10.2).

Cutting

Cutting with an oxyacetylene cutting torch is another way the do-it-yourselfer can bevel the edges of metal prior to welding (Fig. 10.4). This beveling can often be done while the metal is being cut to size. It is important to keep in mind, however, that the cut must be of a high quality to minimize any additional metal edge preparation and to ensure that the pieces fit together well. It is generally accepted that cutting a beveled or chamfered edge on a piece of steel is not something that can be accomplished by a beginning welder. You will therefore be better off to rely on a grinding wheel for your beveling work until you have become quite proficient at both welding and cutting with oxyacetylene.

FIGURE 10.2

Use a bench grinder to bevel the edges of metal.

QUICK»»TIP

One alternative to the conventional grinding wheel is a disc sander or side angle grinder with a special metal cutting abrasive wheel or blade. This hand-held machine can be used for a variety of tasks around the shop. For example, it is handy for cleaning up metal before and after welding. A disc sander is also very useful for smoothing metal prior to finishing with primer and paint (Fig. 10.3).

Welding Steel Plate

For your first exercise in welding thick metal, select two pieces of mild-carbon steel plate, 6 to 8 inches long and about 4 inches wide. Steel plate, in case you are wondering, is metal that is at least $\frac{1}{4}$ inch thick. For this first exercise, you should choose plate that is about $\frac{1}{4}$ inch thick. Steel of this thickness will be more than adequate for learning how to weld heavy metal.

FIGURE 10.3

A hand-held disc sander is very useful for cleaning up metal before and after welding.

FIGURE 10.4

Beveled edges can also be made with a cutting torch. The key is to hold the torch at the right angle and steady.

Before the actual welding can begin, you must first bevel the edges that will butt together to form a joint. As mentioned earlier, the best way to do this is on a bench grinder. Any quality bench grinder will have a guide table in front of each wheel that can be set so that grinding can take place at a predetermined angle. Grind one long edge of both pieces of steel plate to an angle of 45 degrees. It may be helpful at this time to explain how to grind the edge of a piece of steel plate.

Grinding Plate Edges

You should make sure that the bench grinder is securely fastened to a work surface. Good lighting should be over the machine, and you should be wearing some type of clear eye protection. Lightly press one end of the plate into the spinning grinding wheel, and move the piece with a back-and-forth motion. The reason for this is to prevent you from grinding one area more than another. Do not press the metal too hard into the wheel as this will slow down the spinning and reduce the speed at which the metal is removed. During the grinding, it is important that you keep the steel plate flat on the guide in front of the machine (Fig. 10.5). This will ensure that the angle being ground is, in fact, consistent. Stop grinding often, and carefully inspect the edge to make sure it is being ground at both the right angle and uniformly along the edge. Proper grinding takes time, so do not try to rush the work.

Tack Welding Joint Edges

After both edges have been ground, lay the two pieces of steel plate across two firebricks, with the beveled edges aligned. Tack weld both ends of the joint to hold them in alignment while you weld the beveled joint (Fig. 10.6). Use a bit of filler material (welding rod) if you find it necessary for the tack weld.

As a side note, you should have already attached the proper size welding tip into the blowpipe handle. When in doubt as to the proper size welding tip, consult the handbook that came with your welding torch outfit. Here you should also find recommended working pressures for both oxygen and acetylene (see Table 10-1).

Begin welding the beveled joint at the left end, heating the edges of the joint thoroughly before introducing the welding rod. For ¼-inch-thick steel plate, a ³/₃₂- or ⅛-inch-thick, all-purpose welding rod

FIGURE 10.5

Use a guide to help you grind a true bevel on steel plate.

should be adequate. It is important when working with thicker metals to keep the torch flame moving to produce consistent penetration. The size of the molten puddle will be the largest you have ever created, so give the metal time to heat up and flow properly.

Once the puddle is formed at the bottom of the V groove, introduce the tip of the preheated welding rod. Flow the rod into the groove and build up the metal until it has filled the groove and risen to about ¹/₁₆ inch above the joint. Continue along the joint in this manner until you have both filled the V groove and built up the weld bead.

Basically, as previously stated, welding steel plate is quite similar to welding sheet steel. The major difference, of course, is that you must generally work slower and add more filler material to produce a strong weld. By beveling the edges of the joint before welding, you are almost guaranteeing good penetration of the weld. One of the major points to keep in mind is that you must evenly heat up both sides of the joint to the molten stage before adding the filler rod material.

FIGURE 10.6

Tack welding the ends of a beveled edged joint.

Strength Test

After you have welded your first joint in steel plate, you should check the weld for strength. One very good way to do this is to first let the metal cool. Then clamp the work in a vise, and break the joint with repeated blows from a heavy hammer (Fig. 10.7). A sound weld will require a lot of hammering to break while a poor weld will break quite easily. Practice the above welding exercise until you can produce a welded joint that is uniform and strong.

Characteristics of a Good Weld

It may be helpful at this time to describe what a good weld looks like when working with steel plate. The weld bead should be smooth and form an equal junction between the two pieces of steel plate. The width of the bead should be uniform and about 25 percent wider than the metal is thick. For example, if the steel plate is ¼ inch thick, the width of

Table 10-1   Pressure chart for Sears welding tips.

Metal Thickness (inches)	Tip Size (inches)	Welding Rod Diameter (inches)	Oxygen Pressure (psi)	Acetylene Pressure (psi)
$1/32$	1	$1/16$	5	5
$3/64$	2	$1/16$	5	5
$1/16$	3	$1/16$	5	5
$3/32$	4	$3/32$	5	5
$1/8$	5	$3/32$	5	5
$3/16$	6	$3/32$	6	6
$1/4$	7	$1/8$	7	7
$5/16$	8	$5/32$	8	8

the weld bead should be about $5/16$ inch. The underside of the weld should be uniform as well with no excess metal formations. In addition, the bead should be covered with a very thin oxide film on both sides of the joint.

Figure 10.8 shows most of the joint possibilities when welding thick metal. Since these are time-proven joint designs, you should use them whenever possible. All of these joints can be fashioned on a conventional home bench grinder, and the do-it-yourselfer should have little problem in duplicating them.

Multilayer Welding

On very thick metal, over $5/8$ inch, the home welder may find it almost impossible to fill the beveled joint with one pass of the torch. It is standard practice, in the welding industry for the welder to build up the weld bead in several passes. Technically, this is called *multilayer welding*.

FIGURE 10.7

Test the strength of a joint by breaking it in a vise with blows from a hammer. A good weld will require a number of hammer blows while a weak joint will break quite easily.

By approaching heavy metal welding in this manner, the welder will find it much easier to deposit several layers of filler material until the welded joint has been built up to the proper height above the base or parent metal. The point of multilayer welding is to produce smaller puddles of molten metal and enable the welder to have more control over the joint.

When building up a weld bead with multilayer welding, the first layer should provide good penetration at the bottom of the V groove. Additional layers must fuse this first layer with the filler material and the sidewalls of the joint. The final layer seals the joint and should be crowned slightly above the base metal (Fig. 10.9).

Anyone knowledgeable in the art of welding will tell you that you should learn welding in gradual stages, mastering each before

FIGURE 10.8

Common joint designs for metal more than ¼ inch thick.

FIGURE 10.9

Multilayer welding.

moving on. For the home welder, it would be best to begin with thin sheet steel. After mastering welding with and without filler rod, move onto steel plate. Do not attempt to weld plate thicker than ¼ inch until you are able to produce a respectable welded joint. After you have the basics of torch control down, then move on to thicker metal.

Protection and Problems

No welding project can be considered finished until the piece has been pressed into service. In some cases this will mean simply cleaning up the weld to remove the ridges of the weld bead. In other cases, however, this may mean grinding the weld flush with surrounding areas and finishing with several coats of paint or clear finish. This chapter explains protection techniques and examines some common welding problems.

Paint Coatings

The actual use of the welded piece will, of course, determine the extent of the finishing. You should keep in mind that ferrous metals (iron, steel, and alloys of these magnetic metals) will rust if subjected to the forces of the elements. Since rust will in time cause the metal to deteriorate, you will want to prevent this whenever possible.

There are a number of paint coatings that have been specifically developed for use over metal surfaces. Generally speaking, you cannot cover metal with just any paint, such as latex wall paint, and expect this coating to perform in service. You should also keep in mind that the first coat is usually a special primer covering designed to provide a suitable adhesion base for the subsequent finish coats. Probably the best place to find more information about paint coatings is at your local paint store or your welding supply store.

One alternative to a paint coating for welded metals is some type of oil finish. Simply stated, if you coat metals that will be exposed to the elements with oil or some type of silicone spray (such as WD-40), the coating will reduce the chances that the metal will rust. If you do this, keep in mind that the coating must be removed before the metal is ever again heated with a torch flame. Oil or grease burns violently in the presence of pure oxygen, a very dangerous condition, to say the least.

Sandblasting

Before any coating is put on metal, the surface must be cleaned of any existing rust, flux, or scale. Some of the tools that will remove these surface coatings are a wire brush (hand operated or mechanical), sandpaper, or a grinding wheel or other suitable machine. Sandblasting is still another way of cleaning up a metal surface preparatory to finishing. Since sandblasting equipment is expensive, it is recommended that the do-it-yourself welder have this cleaning work done by someone who specializes in the field. A good sandblast cleaning will give a truly professional look to your metalworking project.

Common Welding Difficulties

Undoubtedly the most common problem encountered by beginners is popping of the torch during the welding process. In most cases, this is a result of holding the cone of the flame too close to the molten metal. As you develop some skill at keeping the torch tip at a consistent height above the work, you will find popping less of a problem.

Torch Popping

Another cause of popping, and one that often results in the flame going out, results when the welding tip becomes dirty or clogged from bits of molten metal. The first thing you should do when the flame pops out is to turn off the acetylene and oxygen, in that order. Next, check the tip to see that it is not clogged. If it is, clear it by using a special tip cleaning tool. Use a cleaning wire that is just smaller than the opening. Push the wire in and out of the tip hole several times. It is important that the wire be pushed in and pulled out straight so as not to enlarge or distort the hole in the tip (Fig. 11.1).

You should also remove any accumulation of metal slag around the orifice at the end of the tip.

If you move too quickly along a joint or if the oxygen-acetylene pressures are too low, the bead that will result will be narrow and uneven. When inspecting your practice welds and subsequent welds as well, keep an eye on the quality of the bead. A good weld will resemble those shown in Fig. 11.2.

FIGURE 11.1

Clean your welding tips with a special tip cleaning tool.

Weld Bead Holes

The importance of constant torch movement cannot be overstressed. Holes in a weld bead are caused by holding the cone of the flame in one spot for too long (Fig. 11.3). If holes appear while the metal is still in the molten state, they can usually be filled with a bit of welding rod,

Range of thickness	Number of layers	Method of depositing	Cross section of weld
3/8″ to 5/8″	2	Two layers one pass	
5/8″ to 7/8″	3	Three layers two passes 1 & 2 in one pass layer 3 one pass	
7/8″ to 1 1/8″	4	Four layers three passes 1 & 2 in one pass layers 3 and 4 in two passes	

FIGURE 11.2

Suggested methods for multilayer welding.

FIGURE 11.3

Holes in a weld bead are caused by not moving the blowpipe quickly enough.

FIGURE 11.4

Warped metal can often be hammered back into the desired shape.

by simply touching the rod to the area and letting it flow into the hole.

Strive for perfection in your welding exercises. A good weld bead should be a uniform series of circles with no ridges or high and low spots. This uniformity can come only as a result of a total familiarity with your torch, its adjustment, the filler material, and the metal being joined. Needless to say, it will take you many hours of working with a torch and the various types of metals before you can expect to produce good-looking, as well as mechanically sound, welded joints.

Warping of Metal

Metal has a tendency to warp when heat is applied. This is a problem that is more prevalent with light metal; in short, the thinner the metal, the greater the possibility of some type of warpage. Knowing that metal will expand when heated and contract when cooled can be used to your advantage when working with relatively thin sheet metal. After welding, straightening can be done with your torch. If undesirable warpage presents itself, you can usually correct the problem by hammering and/or heating the opposite side of the warp (Fig. 11.4).

The Weldability of Certain Metals

Y ou already know from reading Chapter 5 about some of the general
characteristics and properties of many metals and their alloys. What
you probably do not know, at this point, is how easy or difficult it is to
weld these different metals. The purpose of this chapter then is to supplement the information in Chapter 5 and let you know what you are
in for when you are faced with welding a specific type of metal.

Iron

Wrought iron is very easy to work with because it approaches the melting point gradually and is therefore easy to control. Wrought iron can
be welded quite satisfactorily with all-purpose steel welding rods.
There are also low-carbon-content steel welding rods that have been
designed specifically for welding wrought iron. You should use this
type whenever possible.

Since cast iron is a chromium alloy, it can be welded very much the
same as mild-carbon steel, with a neutral flame. Actually, there are
two types of cast iron: gray and white. Gray cast iron is the easier of the
two to work with. In just about every instance, it is better to braze weld
cast iron. Square cast iron filler rod, rather than round, is also available for welding cast iron.

Steel

Steel, in one form or another, is probably the most common metal for the do-it-yourself welder. As you know, steel is classified according to its carbon content and/or its alloying element, such as chromium nickel, manganese, or molybdenum.

Mild-carbon steel or low-carbon steel is the easiest type for the home welder to work with. Low-carbon steel is probably the best steel to use when working with the oxyacetylene process. Use mild-carbon steel, all-purpose welding rods for best results.

High-carbon steel can be difficult to weld for the inexperienced hand because you must work quickly and keep the heat as low as practicable. All-purpose mild-carbon steel welding rods can be used for joints in high-carbon steel. Stronger welds can be achieved only by using special high-carbon filler rods, however.

Stainless steel is a difficult metal to weld for the oxyacetylene welder because it contains both nickel and chromium. The best method for welding stainless steel is TIG welding. If you find that you must weld stainless steel, keep in mind that it has a tendency to buckle and warp more than other types of steel. Reduce the chances of this becoming a problem by tack welding at close intervals along the joint before actually beginning the welding. You must weld stainless steel at an even, steady speed.

Low-carbon alloy steels are generally quite easy to weld, especially when working with sheet type alloy steel. These steels generally have a low (0.15 or less) carbon content. An all-purpose mild-carbon steel filler rod works about the best.

Steel that contains a large portion of an alloying element requires that a special welding rod be used for a sound joint. Obviously, the filler rod should contain approximately the same proportion of the alloying element as the parent metal. Your local welding supply house is probably the best source of information about special welding rods.

QUICK»»TIP One professional guide to remember is that if the joint between the unwelded pieces begins to open as you weld, you are moving too slowly. On the other hand, if the joint begins to close in front of the flame, you are moving too quickly.

Copper

There are two types of copper: *electrolytic* and *deoxidized.* The former is difficult to weld because it contains a small amount of oxygen. It is usually brazed with a silver rod rather than welded. Deoxidized copper is easier to weld, providing a special deoxidized copper welding rod is used as a filler material.

Part of the difficulty in welding both types of copper is that the metal does not readily flow until it reaches the molten state. In other words, the metal remains solid until it reaches its melting point, which is about 1,980°F. Once this temperature is reached, the copper will immediately become liquid.

Bronze, as you know, is an alloy of copper. This metal is very easy to weld and is quite popular for metal sculptures. When welding bronze, the obvious filler material to use is bronze brazing rod. You must also use a special flux when welding bronze. The flux forms a protective coating on both the metal and filler material that prevents oxidation of the parent metal during the bonding. The flux also dissolves the oxides that may form and further helps in the flowing of the filler material. Generally speaking, just a tiny bit more oxygen should be used in the flame when welding bronze. This will mean that the flame will be slightly oxidizing. Since the amount of excess oxygen will be influenced by the particular bronze you are working with, a bit of experimentation is necessary when welding this metal.

Aluminum

Aluminum is the most common metal in the world and is also very difficult to weld using the oxyacetylene process. Part of the reason aluminum is hard to weld is that it has a low melting point, about 1,220°F. Unlike other metals, aluminum gives no warning as it approaches the melting point temperature. A special welding process, MIG welding, was developed specifically for working with aluminum because of the problems of welding with the oxyacetylene joining process.

Aluminum, nevertheless, can be welded with oxyacetylene, but it is a chancy proposition. Begin by preheating the metal with an excess acetylene flame. This flame will deposit carbon on the surface and make it look very black. Next, heat the joint area with a neutral flame until the carbon is removed. At this point apply some special aluminum welding flux, which is liquid. Then, while keeping the cone of the flame at least 1 inch above the joint, introduce a special aluminum welding rod into the flame. The object here is to tin the joint with the aluminum rod. After the joint has been tinned, melt more rod onto the area and build up the joint. It is important that the flame of the torch not be played on the joint as this may cause the aluminum to become very weak. After the weld has been completed, all flux must be removed from the joint and surrounding area. This is commonly done with warm water and a brush.

There is no special talent required to learn how to weld. What is required, however, is a determination to master the basics of flame control and a great deal of patience. You should practice on light sheet steel in the beginning and not move onto thicker metals until you have mastered the basics of torch and flame control. In time you will develop a certain sense about how metals are affected by intense heat, how metals flow, and just how much heat is necessary to manipulate molten metal. Practice until you know the characteristics of one metal; then expand your knowledge by experimenting with other types of metal.

Cutting Metal

Undoubtedly, the most dramatic use of oxyacetylene equipment is in cutting metal. Sparks shower from the work, and, in most cases, the cutting progresses very quickly. The equipment necessary for cutting metal with the oxyacetylene process is basically the same as for welding. The major difference, of course, is that the blowpipe must be fitted with a special cutting attachment. In addition, the working pressures are generally greater than for standard welding (Fig. 13.1).

Equipment

In most cases, a special cutting attachment is fastened onto the end of the standard welding blowpipe handle in place of the welding tip. It will be to your advantage to understand the principles behind its operation (Fig. 13.2).

Cutting Attachments

Basically, all cutting attachments are the same, although design features often differ between manufacturers. A cutting attachment will have a housing through which three tubes run. One tube carries acetylene gas to the tip. Another similar size tube will carry oxygen to the tip. A third, larger tube will also carry oxygen to the tip. The first two tubes carry oxygen or acetylene to a special expansion chamber just

FIGURE 13.1

Cutting in progress.

FIGURE 13.2

A typical welding torch with cutting attachment.

behind the cutting tip. There the gases are allowed to expand and mix thoroughly before passing through the small holes in the cutting tip.

In most cases there will be six holes in a cutting tip, surrounding a larger hole in the center. The third tube in the cutting attachment handle carries pure oxygen, and it leads directly to the large center hole. The flow of pure oxygen through the tube to the center hole is controlled and regulated by a lever that sits on top of the cutting attachment handle. No oxygen can flow through this tube—or out of the large center hole—until the lever is depressed (Fig. 13.3).

Cutting Tip Holes

The six small holes in the cutting tip are called *preheat holes,* and the larger, center hole is called the *cutting hole.* Each of the preheat holes produces a flame that resembles a standard welding flame. Since the oxyacetylene mixture flows equally through all of these small preheat holes, each flame will be the same: acetylene, carburizing, neutral, or oxidizing. As you can well imagine, the six preheat holes produce an enormous amount of heat (Fig. 13.4).

The center hole in the cutting torch tip does not produce any flame. Instead, when the lever is pressed, a stream of pure oxygen is introduced into the heat zone created by the preheat flame. This causes the metal to oxidize away very rapidly. It may be helpful at this time to explain how the oxyacetylene cutting process works.

Chemical Reaction

Oxyacetylene flame cutting can be used for many common ferrous metals. Oxyacetylene cutting can successfully be used to cut low-carbon steel, low-carbon alloy steel, wrought iron, and other ferrous metals.

FIGURE 13.3A

A cross-section view and parts for Airco Series 9500 hand cutting.

831 2282 valve cap

831 0850 valve & seat ass'y

831 2274 lift pin

831 2283 lift pin bushing

Enlarged section A-A

831 8030 valve spring

831 8255 packing ring

831 2290 seat holder extension

830 1316 "O" ring (¼" O.D.)

831 2291 oxygen nipple

803 0008 hose connection gland

803 0003 fuel gas hose connection nut

803 0004 oxygen hose connection nut

831 2234 fuel gas valve ass'y.

831 2242 fuel gas valve body

831 2124 valve stem ass'y.

Not furnished with torches

9500 series torches

831 2147 packing only

831 2270 rear end

831 2273 preheating valve stem ass'y.

No. 10-32 fil hd. brass mach screw ⁵⁄₁₆" lg.

831 2255 packing only

(2) No. 10 light spring lockwasher steel cad. plated.

831 2264 lever

831 2281 handle

831 2265 front end

(4)No. 4-40 fil. hd. stainless steel mach. screw ⅛" lg.

831 2266 lever pin

8 31 2268 yoke

No. 8-32 headless cup point stainless steel set screw ¼" lg.

Straight head & tubes ass'y.

45° head & tubes ass'y.

75° head & tubes ass'y.

90° head & tubes ass'y.

Head

Cutting oxygen tube

Fuel gas tube

Preheat oxygen tube

831 2259 tip nut

165

FIGURE 13.3B

A cross-section view and parts for Airco Series 9500 hand cutting torches.

This process does not work well for cutting stainless steel, cast iron, or any nonferrous metal such as aluminum.

Actually, the process of flame cutting is a chemical reaction. This reaction takes place because oxygen has a chemical attraction to ferrous metals when they are heated above their melting point. When excess pure oxygen is added to red-hot ferrous metal, the iron will oxidize very rapidly. To say it another way, the metal will burn up. The force of the excess oxygen, in addition to burning up the metal, will also aid in blowing away the molten metal.

Success Factors

There are several factors that will influence the success of any oxyacetylene cutting process. It is important to use the proper size cutting tip for the particular cutting project at hand. All makers of oxyacetylene torches and cutting attachments offer a selection of various size cutting tips, each designed for a specific thickness of metal. Obviously, you should use the tip size recommended for any given thickness of metal.

The thickness of the metal being cut, in addition to dictating the cutting tip size, will also determine the speed at which the metal can be cut. It is important to keep in mind that the metal is heated up to the "cherry red" stage before additional oxygen is introduced. Obviously, thin metal will reach this stage much more rapidly than thicker metal. It is important, then, from the standpoint of an acceptable cut, for the welder to keep a sharp eye on the area being preheated and to watch for the metal to reach the proper temperature, based on the color of the metal. The preheat flames are kept operating during the entire cutting operation. The oxygen is introduced *only* when the metal has reached the proper temperature. If the excess oxygen is added before the metal has reached the proper temperature, oxidation or cutting will not take place.

As with welding, the best way to learn how to cut metal with oxyacetylene equipment is to practice often and on various thicknesses of metal until you have mastered some of the basics. It will be helpful, therefore, to run through several exercises in flame cutting and point out the important functions that must be accomplished during the cutting operation. Before we begin, however, it may be helpful to quickly run through how to set up the oxyacetylene cutting equipment.

1. Cutting jet oxygen passage
2. Preheating-flame passages

FIGURE 13.4

A cross-section view of an acetylene cutting tip.

> **QUICK»TIP**
> Each cutting tip will require specific working pressures for both oxygen and acetylene. In all cases the working pressure for oxygen will be considerably more than the working pressure for acetylene, as Table 13-1 illustrates.

Setup

Assuming that your regulators have already been attached to both the oxygen and acetylene cylinders and that the standard welding torch handle is also attached to hoses running from the regulators, begin by removing the welding tip attachment. In most cases this will entail unscrewing the welding tip holder from the end of the handle. At this time, the tanks should be turned off and the lines cleared of both acetylene and oxygen (Fig. 13.5).

Table 13-1   Pressure chart for Sears cutting tips.

Plate Thickness (inches)	Tip Size (inches)	Oxygen Pressure (psi)	Acetylene Pressure (psi)
3/8 − 5/8	0	30 − 40	5 − 15
5/8 − 1	1	35 − 50	5 − 15
1 − 2	2	40 − 55	5 − 15
2 − 3	3	45 − 60	5 − 15
3 − 6	4	50 − 100	5 − 15

FIGURE 13.5

The cutting attachment fits onto the blowpipe handle.

FIGURE 13.6

The proper size cutting tip is fitted onto the cutting attachment.

The cutting attachment is then fastened to the end of the blowpipe handle and secured. The next step is to fit the cutting attachment with the proper size cutting tip for the project at hand. For our first exercise, we will be cutting light gauge steel, less than 1/4 inch thick. Your welding outfit should have come with a booklet with recommendations for cutting tip sizes for cutting various thicknesses of metals. Choose the right size tip, and fit it into the end of the cutting torch handle (Fig. 13.6). In all probability, you will be attaching a relatively small cutting tip.

Oxygen Flow

Once the proper size tip has been attached and you have double checked all connections to make certain that they are snug, open the oxygen tank valve slowly and allow some of the oxygen to pass into the regulator housing. At this time the high-pressure gauge of the regulator should register how much pressure is in the oxygen tank. Open the valve slowly so that a sudden surge of pressure will not damage the internal parts of the regulator. The diaphragm of the regulator should be wide open at this time. The low-pressure or working-pressure gauge should read zero. The control valves on the torch and cutting attachment should also be closed at this time as well. The next step is to set the working pressure for oxygen. This is done by simply turning the oxygen regulator control lever until the low-pressure gauge registers. Set the working pressure according to the recommendations for the cutting tip being used. This information can be found in the manual that came with your welding outfit (Fig. 13.7).

The next step is to give a quick check to all oxygen connections to make sure that no oxygen is leaking into the atmosphere. If all is well, open the oxygen control valve on the welding torch handle two full turns. At this time no oxygen should flow. The reason for this is that the oxygen flow is actually controlled by the oxygen control valve on the cutting torch handle rather than the valve on the welding torch itself (Figs. 13.8 and 13.9).

Open the oxygen valve on the cutting torch handle a quarter turn to make sure oxygen flows. Then turn it off quickly. Next depress the oxygen cutting lever for a second, and then release it quickly. When you depress the oxygen cutting lever, a stream of oxygen should flow from the center hole in the cutting tip.

After the oxygen system has been pressurized and checked for leaks and proper operation, you can turn your attention to the acetylene line. First make sure that the acetylene control knob on the torch handle is in the off position and the regulator control lever is also turned off (fully open). Open the acetylene tank valve slightly to let some of the fuel into the regulator housing. The high-pressure gauge should register how much pressure is in the acetylene tank (Fig. 13.10).

Acetylene Flow

The next step is to turn the acetylene regulator lever until the working-pressure gauge registers the proper pressure. Once again, this information can be found in the booklet supplied with your welding outfit. After the proper working pressure has been set, open the acetylene control knob on the torch handle to make sure

FIGURE 13.7

Open the valve slowly so that the regulator is not damaged by a sudden surge of pressure.

FIGURE 13.8

The oxygen control knob on the torch handle is opened all the way, but the flow of oxygen is controlled by the oxygen control knob on the cutting attachment.

Oxygen control lever

Oxygen control knobs

FIGURE 13.9

The oxygen control knob on the cutting attachment.

FIGURE 13.10

Set the working pressure for oxygen.

acetylene is flowing (Fig. 13.11). Then quickly turn it off. Give a quick check to the acetylene connections to make certain that there are no leaks.

As you can see, the only difference between setting up the oxygen and acetylene lines is that there are two control knobs for oxygen, one on the torch handle and another on the cutting attachment. There is only one for acetylene, on the welding torch handle. This may cause a bit of confusion initially, but I plan to make it clear as to which oxygen control knob is to be used when cutting.

The flow of acetylene is controlled by one knob on the torch handle. The oxygen knob, next to this, is opened all the way when cutting. It is important to remember that no oxygen will flow out of the torch cutting tip, however, until the oxygen flow control knob valve on the cutting attachment is opened. To summarize, the flow of acetylene is controlled by opening the acetylene knob on the torch handle. The oxygen control knob on the torch handle is opened at all times, but the flow of oxygen is regulated by the oxygen control knob on the cutting handle attachment (Fig. 13.12).

Precutting Procedures

The next step in learning some of the basics of oxyacetylene cutting is to lay your hands on a piece of scrap steel plate, about ¼ inch thick, 4 inches wide, and 8 inches long. Lay this scrap lengthwise between two firebricks on your worktable. Now, with protective goggles in place, open the acetylene control valve a quarter turn, and light the torch. Then open the oxygen valve on the cutting attachment. The valve on the torch handle next to the acetylene control knob is, of course, open already. Develop a neutral flame at the preheat holes that results in a cone about ⅛ inch long. You will find this easiest to accomplish by alternately increasing the flow of acetylene and then oxygen until the proper flame is achieved. When you are done adjusting to a neutral flame, each of the preheat holes should have this flame (Fig. 13.13).

After the neutral flame has been developed, depress the oxygen cutting lever momentarily to make certain that oxygen is, in fact, flowing through this part of the cutting attachment. You may find, when you depress the cutting lever, that the preheat flame needs to be adjusted. This is often the result of excess oxygen in the center of the preheat flames. Adjust the preheat flames if necessary. Now the torch is ready for the cutting operation.

FIGURE 13.11

Set the working pressure for acetylene.

FIGURE 13.12

The flow of oxygen is controlled by this knob on the cutting attachment.

Positioning of Hoses

Before actually beginning to cut, you should take a quick look around the work area to make sure the oxygen and acetylene hoses do not run under the work. This would be a dangerous condition especially if a spark or hot slag were to fall onto the hoses. It might cause the hose to

FIGURE 13.13

Adjusting to a neutral flame.

rupture, and a fire could result. For the record, it is always best for the hose to run from behind the welder. That way your body will always shield the hoses from hot sparks and molten slag (Fig. 13.14).

You should also look to be sure you have a clear path to the oxygen and acetylene tanks. If a fire were to break out, turn off the flow of oxygen and acetylene at the tanks after turning off the control valves on the torch handle and cutting attachment.

Access to Safety Equipment

Safety equipment should also be within easy reach. This includes a fire extinguisher. See Chapter 14 for solid information on firefighting equipment. A bucket of sand and a bucket of water should also be in the welding shop for a quick response to any type of fire (Fig. 13.15).

FIGURE 13.14

Oxygen and acetylene hoses should come from behind the cutter.

As mentioned earlier in this chapter, you should have tinted goggles over your eyes. Many experts suggest that the molten metal being cut is brighter than molten metal being welded. Make sure your goggles are dark enough to protect your eyes while at the same time allowing a clear view of the work in progress (Fig. 13.16). Heavy leather gloves and other body protection should also be worn as sparks and hot slag are a common occurrence when cutting metal with an oxyacetylene cutting unit. It is important to wear the proper clothing when cutting or welding. Chapter 14 contains extensive information about the types of clothing that are suitable to wear when working in the welding shop.

FIGURE 13.15

Adjusting to a neutral flame. A fire station should be within easy reach of the cutting area.

Cutting a Hole in Steel Plate

To get a feel for how a cutting torch can pierce through metal, the first exercise will involve cutting a single hole in the center of the steel plate. Begin by adjusting the torch for a neutral flame, then add a bit more oxygen to develop a slightly oxidizing flame—which is better for cutting than a neutral flame. Hold the tip of the torch about $\frac{1}{8}$ to $\frac{1}{4}$ inch above the center of the steel plate. The idea is to hold the white cones in each of the preheat flames just above the surface of the metal. In a matter of moments, you will notice the change in the color of the metal until finally the color is a cherry red. At this time, depress the oxygen cutting lever on the cutting attachment. It is important to press the lever slowly to blow a stream of pure oxygen onto the molten steel.

FIGURE 13.16

Tinted welding goggles should always be worn when working with a cutting or welding torch.

As the oxygen hits the molten metal, a shower of sparks will appear at the center of the cut. If you hold the torch steady, the oxygen will remove the surface layers of the metal very quickly, exposing the lower layers. These layers will almost immediately be heated by the preheat flames, and preoxygen will blow them away as well. In a very short time, you will blow a hole in the center of the metal. This basically is how all oxyacetylene cutting is done. The difference between blowing a single hole through steel and cutting a straight line across a 2-foot piece of steel is simply a matter of practice and finesse (Fig. 13.17).

Some general comments about cutting are in order at this time. It is important that the cutting torch be held at a constant and steady height above the metal surface. You will find it helpful, especially in the beginning, to use two hands to hold the torch (Fig. 13.18). Assuming that you are right handed, your right hand will hold the cutting torch at the balance point, and your right thumb will be on top of the oxygen cutting lever. Your left hand will be in front of your right and will serve as a

QUICK>>>TIP One last reason for holding the cutting torch ⅛ to ¼ inch off the surface is that at this height the preheat holes and oxygen cutting hole will be protected, somewhat, from reflected heat. At this height, the holes are less likely to become clogged with flying bits of molten metal.

FIGURE 13.17

Just before the white-hot stage, press the oxygen cutting lever.

steadying point for the cutting torch. Of course, both hands should be encased in heavy leather gloves to protect them from hot sparks and slag.

Cutting a Line across Steel Plate

The next exercise involves cutting a straight line across the steel plate. If you did not burn too many holes in the piece of steel you used for the first exercise, you can use it for this exercise as well. Before cutting, you must mark the steel plate so you will have a guide for the cutting. The accepted method of marking metal in the welding trade is to use a special *soapstone pencil* (Fig. 13.19).

Soapstone is simply a type of talc that remains visible under heat. It is used by welders for marking straight and curved cut lines. Special pencils are available that can be refilled with lengths of new soapstone, and they work very much the same as a common mechanical lead pencil.

One alternative to using a soapstone pencil to mark cuts on metal is to mark the cut line with a series of indentations spaced about ¼ inch apart. These are made with a center punch and a hammer.

FIGURE 13.18

Use two hands to hold the cutting torch and to help you cut in a straight line.

FIGURE 13.19

A soapstone pencil is the best tool for marking metal to be cut.

While this alternative method works well, it will take you quite a bit longer to mark your cuts than if you simply use a soapstone pencil (Fig. 13.20).

Mark the scrap steel plate with a line running across the piece (Fig. 13.21). Next position the steel plate so that the cut line is about midway between the two firebricks holding up the steel plate. Then light the torch, develop a neutral flame, and begin heating the metal on one edge of the cut line.

FIGURE 13.20

A punch and a hammer can also be used to mark a cut line on metal, but this will take longer than using a soapstone pencil.

FIGURE 13.21

The cut line on a piece of steel plate.

Depressing the Oxygen Lever

Once the edge of the steel plate is cherry red in color, the cutting oxygen lever is slowly depressed. This will allow a stream of pure oxygen to blast against the molten metal, resulting in sparks and the beginning of a gouge or cut.

Depressing the oxygen lever slowly prevents the heated spot on the metal from being cooled by the flow of oxygen. This cooling would obviously prevent cutting from taking place (Fig. 13.22).

After the metal has been gouged by a fine stream of oxygen, you will notice that the hole will become deeper. It is at this time that you should press the oxygen lever once again to complete the cut. At about the same time, you should begin to move the cutting torch forward along the cut line about ½ inch. Since this metal is very close to that which has just been severed with oxygen, it will be very hot. You will have to heat it with the preheat flames only momentarily. Then spray a little of the pure oxygen onto the area. This will cut the metal. As this happens, move the torch flames forward once again. Basically, this is how all metal is cut with an oxyacetylene flame and a stream of pure oxygen.

Factors Influencing Speed

The speed at which you progress along the cut line will be determined by a number of factors, the thickness of the metal being the most crucial. A slow even motion is best when cutting metal with oxyacetylene equipment. It will take time to learn just how much heat is needed before pressing the oxygen cutting lever.

Proceed in the manner described until you have made one pass over the steel plate, along the soapstone cut line. Turn off your cutting torch, and let the metal cool until it can be inspected. If you are in a hurry, you can drop the pieces into a bucket of water to cool them rapidly (Fig. 13.23). If you do this, it is a good idea to pick the hot metal up with a pair of pliers. I always keep a pair on the welding table for such tasks. When I am working off the table, I often stick a pair of pliers in my back pocket for instant use.

After the metal has cooled, look closely at your first cut. There are basically four possibilities as to how the cut will appear: perfect, too slow, too fast, or too much heat. Since each of these cuts has definite characteristics, it will no doubt be helpful to discuss each of them.

Perfect Cut

A perfect cut will, at first glance, appear as if the metal has been cut with a hacksaw. There will be no, or very little, buildup of metal on either the top or bottom edge of the cut. The face of the cut will appear to have fairly straight lines running through the metal, and each will be about even with surrounding marks as far as depth. In short, a good cut made with an oxyacetylene cutting torch will look good, almost as if it were made with a machine (Fig. 13.24).

Too Fast Cut

A cut through steel plate that has been made too quickly will have a buildup of

FIGURE 13.22

The oxygen cutting lever should always be pressed slowly and then held when the proper cutting action is reached.

FIGURE 13.23

Metal can be quickly cooled in a bucket of water.

FIGURE 13.24

A good-quality cut with oxyacetylene cutting equipment.

FIGURE 13.25

The cutting speed was too fast.

FIGURE 13.26

The cutting speed was too slow.

FIGURE 13.27

The preheat was too hot.

slag-like metal on the bottom side of the plate. This is a result of not allowing enough time for the pure oxygen to blow the slag out and away from the metal. The cut lines on a piece of metal that has this slag buildup will be curved, almost as if the metal were cut with a circular saw (Fig. 13.25).

Too Slow Cut

In all probability, your first cut through steel with oxyacetylene will have a buildup of slag on the top, the bottom, and on the face of the cut. You may find that in some areas the cut lines are straight, while in other areas there are no cut lines. This irregularity is due largely to heating the metal too much. In other words, you could have moved the cutting torch more quickly along the cut line.

Another characteristic of moving the cutting torch too slowly may be burn holes in the top of the cut. If slow movement is combined with an irregular torch movement, you could end up with quite an ugly edge to the cut metal. This will appear as bubble-like mounds and very uneven cut lines on the face of the metal (Fig. 13.26).

Too Much Preheat Cut

When more heat than is required is applied to the cut line before depressing the oxygen cutting lever, the top of the metal will indent. This happens because more of the metal along the cut line is ready for the chemical reaction of the oxygen. In extreme cases of too much preheat, the top edge of the cut will appear to flow down the face of the cut. In effect, this is exactly what has happened (Fig. 13.27).

Professional Cutting Tips

As you can see from the description of the various cuts, the speed, height, and intensity of the cutting torch preheat flames are points over which you should have the most possible control. It will be helpful to mention a few professional tips to help you make good clean cuts in metal with an oxyacetylene cutting torch.

Observe the Metal

One of the most important things that can be learned by beginning welders (in this case beginning oxyacetylene cutters) is to keep a very close eye on the metal. As it heats up to about the right temperature, you should be ready to depress the oxygen cutting lever and introduce excess oxygen to the cut area. The novice is really at a disadvantage here even if he or she has a keen eye. Until you have watched metal heat up with an oxyacetylene flame a few times, you will not really know what to look for. You must be aware that the metal will move up in temperature quite rapidly. It will change from gray to red quickly and then pass through several shades of red until it is practically white in color. If heat is constantly applied, the metal will turn white and become truly molten, just as when you were puddling metal during the welding practice exercises.

When the metal has become molten, without first adding pure oxygen, it will not cut well. When the oxygen is introduced at this point, the metal will deteriorate very rapidly. The time to depress the oxygen cutting lever is just before the white-hot stage (Fig. 13.28). Only then can you expect to make clean cuts in metal.

FIGURE 13.28

You must depress the oxygen cutting lever just before the metal reaches the white-hot stage.

Adding Oxygen

The manner in which you depress the oxygen cutting lever will have a very real effect on how the metal cuts and even on how the metal will look after being cut. It is a common fault of beginners to add too much oxygen when cutting. This is understandable but not acceptable.

The proper way to add oxygen to cherry red metal is to very slowly press on the oxygen cutting lever until the metal is flowing away at an even stream. This is actually a very soft, subtle touch that can be developed only over time. In time, however, you will learn that only a bit of oxygen is needed and then you will develop the soft touch necessary to make a good, clean cut in any thickness of metal.

Proper Torch Movement

Keeping the cutting torch moving along a straight line, while at the same time keeping the preheat flames at the same height above the metal, is another requirement for a good cut in metal. As I mentioned earlier, you should use your left hand as a steadying point for the torch while controlling the oxygen cut lever with your right hand and thumb. This is an awkward position when you have to move the torch to cut along a straight line. There will be a natural tendency to put a curve in the cut. One way to overcome this is to move your left forearm and right hand along the line. The task will be easier to accomplish if you also move your whole body at the same time. A good work table will help you to move in front of the work while cutting with oxyacetylene (Fig. 13.29).

One way to eliminate much of the body movement necessary to make a straight cut is to attach a temporary guide for the cutting torch. Then you can simply rest the barrel of the cutting torch on the guide. Move the torch along the cut line at both the right height and along a straight path.

A guide for any oxyacetylene cutting torch can be fashioned quite simply from pieces of scrap steel you should have

FIGURE 13.29

For a straight cut, lock your elbows against your sides, and move your whole body.

around the welding shop, like a piece of angle iron (Fig. 13.30). The only requirements of the metal used for the guide are that it must be straight, at least as long as the width of the cut being made, and, probably most important of all, the right height for the torch being used.

Preheat Flame Height

As you are probably aware, the preheat flames should be at a height of about ⅛ to ¼ inch above the surface of the cut line. If the white cones are too high above the work, the metal will not heat properly or will take too long to do so. On the other hand, if the heat cones are too close to the work, you will heat up the surface of the metal too much, along with possibly clogging the holes on the face of the cutting tip.

Setting up a Guide

Obviously some experimentation is called for on your part when setting up a guide for your cutting torch. Once you have found an angle iron that is a suitable height for the torch and tip you are using, you should be able to obtain good results.

The guide, if it is angle iron, can be quickly fastened in place parallel to the

FIGURE 13.30

A guide for cutting.

FIGURE 13.31

The cutting guide is simply clamped along the cut line with a pair of Vise-Grips.

cut line with clamps or locking pliers (Fig. 13.31). When positioning the guide, do not place it too close to the area to be cut or the clamp could be damaged or fused to the workpiece. On the other hand, do not place the clamp too far away from the cut line or it will be ineffectual. Most experts agree that a distance of approximately ¾ inch is about the best distance for standard cutting with oxyacetylene.

A guide for an oxyacetylene cutting torch will enable you to make straight cuts on metal. But you still must keep a close eye on the metal so that you will know when to depress the cutting lever. You must also proceed along the cut line at a speed that is adequate for the thickness of the metal being cut.

You should practice cutting straight lines in ¼-inch-thick steel plate until you can make perfect cuts every time. In the beginning, use some type of guide for the torch. This will enable you to develop a cutting technique with greater ease. After you have mastered this, try cutting without the aid of a guide for the torch. In time you should become proficient at cutting freehand.

Cutting Thin Metal

Thin metal can also be cut with an oxyacetylene cutting torch. Thin metal is any type that is less than ¼ inch thick and is most commonly referred to as *sheet metal.* For cutting sheet metal, a different approach is needed as there is always a tendency to burn holes in the metal. Two professional pointers will be of use to you when cutting thin metal.

Hold the torch at an angle to the cut line on the surface. A 45-degree angle, or less, will enable you to make the cut without leaving large holes in the cut line (Fig. 13.32). Another point to keep in mind is to work quickly. The more time you spend heating the thin metal, the greater the

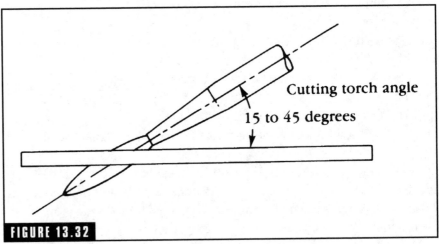

Cutting torch angle
15 to 45 degrees

FIGURE 13.32

Cutting thin sheet metal with an oxyacetylene cutting torch. The torch angle is slight to moderate.

chances of warping the metal or burning holes rather than cutting a straight line.

When cutting sheet metal, you should use the smallest possible cutting tip. Some experts can make cuts in thin metal with a large welding tip and a slightly oxidizing flame (Fig. 13.33). In time, you may want to experiment with cutting with a welding tip also. For now, however, use a small cutting tip.

Bevel Cutting

As you may recall, steel that is thicker than ¼ inch must have beveled edges or it will be difficult to weld with the oxyacetylene process. In Chapter 10 directions were given for grinding the edges of metal to develop a bevel. You may have already tried this and discovered that it takes quite a bit of grinding to bevel the edges of steel plate. Another way to bevel edges of plate steel is, of course, to cut a bevel with a cutting torch (Fig. 13.34). While this type of cutting is not difficult, it does require a slightly different approach than that used for straight cutting.

The major difference between straight cutting of steel plate and bevel cutting of the same material is the angle at which the cutting torch is held. To cut a bevel, the torch tip must be held at a 45-degree angle to the cut line rather than perpendicular to it (Fig. 13.35). In most cases, some type of guide for the cutting torch will enable you to cut at both a constant bevel and along a straight line.

FIGURE 13.33

Light sheet steel can be cut with a welding torch. A very slight angle is used for the torch.

Cutting torch

Steel plate

Torch angle 45°

FIGURE 13.34

Cutting a bevel edge on steel plate with an oxyacetylene cutting torch.

FIGURE 13.35

The cutting torch is held at a 45-degree angle to the cut line to make a bevel cut.

QUICK»»TIP Use a slightly oxidizing flame when cutting a bevel on steel plate. To refresh your memory, first develop a neutral flame for the preheat holes, and then add a bit more oxygen to the flames.

It is important, when making a bevel cut, to keep the torch moving at a consistent speed. An unsteady cutting speed will often result in an irregular cut and possibly a stop in the cutting action.

Cutting Thick Steel

Cutting thick steel can also be done with an oxyacetylene cutting torch. Generally speaking, the thicker the metal, the longer it will take to make a cut. The cutting operation is basically the same as those previously discussed with a few modifications. One of the first things to keep in mind is that the cutting torch tip must be held closer to the work surface of the metal. In most cases this will mean $\frac{1}{16}$ to $\frac{1}{8}$ inch above the cut line. It is also important that the torch tip be held at a right angle to the cut line at all times. This will ensure that the heat is being applied evenly as the cut progresses (Fig. 13.36).

Probably the most difficult part about cutting thick metal with oxyacetylene is starting the cut. There are several different ways of starting a cut that are used by professional welders.

One method of starting, assuming that you are right handed, is to begin at the edge of the metal and move from left to right. This will give

1. Start to preheat; point tip at angle on edge of plate

2. Rotate tip to upright position

3. Press oxy valve slowly as cut starts, rotate tip backward slightly

4. Now rotate to upright position without moving tip forward

5. Rotate tip more to point slightly in direction of cut

6. Advance as fast as good cutting action will permit

7. Do not jerk; maintain slightly leading angle toward direction

8. Slow down; let cutting stream sever corner edge at bottom

9. Continue steady forward motion until tip has cleared end

FIGURE 13.36

Recommended procedure for efficient flame cutting of steel plate.

QUICK»TIP **Some professional welders like to chip the edge of the cut line with a cold chisel and hammer before starting the cutting operation (Fig. 13.37). The sharp edges of the chip will preheat and oxidize more rapidly after doing this.**

FIGURE 13.37

Some professionals like to chip the edge of thick steel plate before starting the cut.

you a clear view of the cut and let you look into it to make sure the slag is being blown out by the oxygen. When moving from left to right, you may have some difficulty in following the cut line. You can usually overcome this by using some type of guide for the cutting tip.

Another method used to start a cut in thick metal is to begin at the corner of the metal by slanting the torch in a direction opposite the direction of travel. As the corner heats up to cherry red and is cut, the welder then moves the torch to a vertical position until the steel has been cut. Then proceed along the cut line.

One last method of starting a cut involves using an iron filler rod. The rod is placed along the cut line before starting (Fig. 13.38). Then, as the preheat flames are applied, the filler rod will reach the cherry red stage very quickly. As soon as this happens, the cutting oxygen lever is depressed causing the filler rod to oxidize. A chain reaction then takes place, and the parent metal begins to oxidize at about the same time.

Whichever means you use for starting a cut, and you should try them all, it is important to move the cutting torch at a steady rate once you begin the cut. After the edge of the metal has been cut, the oxygen cutting lever should be kept constantly open by pressing it lightly. As the metal heats up along the cut line, more cutting oxygen is added. At the same time move the torch slowly and steadily along the cut line.

Pipe Cutting

One last use of the oxyacetylene cutting torch that we will discuss is cutting pipe. Since pipe differs dramatically from plate—it is after all round—a different approach is required.

Pipe Marking

Before metal pipe can be cut, it must be marked with a soapstone pencil and a special marking guide. Any welding supply house will sell special pipe marking paper, but you can usually find something around the shop that will do the job just as well and, it won't cost you anything. What you need is a strip of stiff paper, such as matboard, about 1 to 2 feet long and about 6 inches wide (Fig. 13.39). This paper is then wrapped around the pipe until it overlaps itself. Make sure the edges overlap evenly and then mark the pipe for cutting.

After the pipe has been clearly marked for cutting, place it on your worktable. Prevent the pipe from rolling off the table by placing it between firebricks (Fig. 13.40). The area to be cut should extend off the worktable.

Torch Angle

The size of the pipe will determine for you the angle at which you hold the cutting torch. For pipe that is less than 4 inches in diameter, you should hold the torch so that the preheat flames are almost tangent to the circumference of the pipe (Fig. 13.41). For pipe that is thicker than 4 inches, you can usually hold the torch perpendicular to the pipe without burning through the other side of the pipe (Fig. 13.42).

The thickness of the walls of the pipe, rather than the diameter of the pipe, will determine the size cutting tip to use in the torch. Consult the guidebook that came with your torch for the proper tip size recommendations for the pipe you are cutting.

Two-Person Operation

Begin cutting anywhere along the cut line on the pipe. As the metal reaches the cherry red stage, press the oxygen cutting

FIGURE 13.38

A welding rod is used here as a cutting guide and aid.

FIGURE 13.39

A piece of thick paper or cardboard, with a straight edge, can be used as a marking guide for pipe.

FIGURE 13.40

Steady a length of pipe between two bricks before cutting.

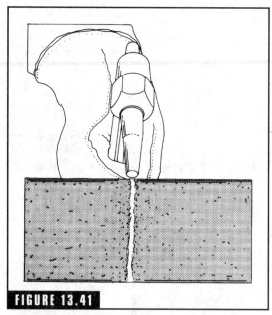

FIGURE 13.41

Cutting small pipe less than 4 inches in diameter requires only a slight torch angle.

lever slowly until you get a good cutting action going. As the actual cutting begins, you must either move the torch along the cut line or have someone else turn the pipe for you. In most instances, cutting pipe will be easier if it is a two-person operation. One person does the actual cutting, and the other rotates the pipe, as required. Needless to say, both welder and turner should wear suitable protective clothing and goggles.

Beveling Pipe Edges

Beveled edges are also possible when cutting pipe. You will find the work a bit easier to accomplish if you point the torch toward the short end of the pipe as you proceed with the cutting (Fig. 13.43).

If you find that you must work alone when cutting pipe, and the pipe is manageable by one person, you can do the cutting on the ground and use your foot as a means of holding the pipe (Fig. 13.44). Needless to say, you should exercise extreme caution when doing this type of cutting. It is almost as much a balancing act as a cutting operation. Heavy work boots are also required.

Piercing a Hole through Steel Plate

The first exercise in this chapter involved burning a hole in the center of a piece of steel plate. At some point, you may find that you require a hole in a piece of metal. It will be to your distinct advantage to know how to approach the task. The process is really quite simple and can be mastered in a short time. Begin by choosing a cutting tip that is one size larger than that which is recommended for the standard cutting on the particular thickness of metal you are working-

ing on. Next, hold the lighted tip of the cutting torch ¹⁄₁₆ to ¹⁄₈ inch above the surface of the metal. It is important that the tip be held at a perpendicular angle.

Heat the spot until it is cherry red, and then very slowly depress the oxygen cutting lever. At the same time, the nozzle of the cutting torch should be raised slightly off the surface, say, ¹⁄₂ inch. This will reduce the chances of slag being blown up into the nozzle holes and clogging them. As the cutting takes place, you may find it necessary to lower the tip slightly and then raise it once again. In a matter of moments, a hole should be blown through the plate. When this happens, lift the torch up and away from the work.

When piercing a hole in steel plate, there is always a risk of clogging the holes in the cutting tip. You should therefore keep a close eye on the flames. If they appear erratic, chances are very good that bits of molten metal or slag are building up on and around both the preheat and oxygen cutting holes. If this happens, you should stop work and clean the tip with a special tip cleaning tool (Fig. 13.45).

Oxyacetylene cutting is a dramatic experience because of the volume of sparks produced and the speed at which metal can be severed. It will take several hours of practice before you can cut metal well, but it is a skill worth perfecting. There is something almost magical about working with molten metal, even though there are inherent dangers.

FIGURE 13.42

Pipe larger than 4 inches is cut with the torch held at a right angle to the cut line.

FIGURE 13.43

The torch position for making a bevel cut on pipe. Point the torch toward the short end of the pipe.

In the next chapter we will discuss how to safely accomplish many welding and cutting tasks. It will be to your distinct advantage to make safe welding and cutting the only way you know how to work.

FIGURE 13.44

If you are very careful, you can cut pipe in this fashion.

FIGURE 13.45

A clean cutting tip works best.

Safety

From the moment we enter into this world until the last beat of our heart, there will always be some type of danger. Depending on how and where we live our lives, the way we get from one place to another, and how we accomplish various tasks, we always run the risk of injury or worse. That is the natural scheme of things.

The home welder, in addition to all of the normal risks to health, lives just a bit closer to danger simply because of the nature of welding equipment. Any welding or cutting operation should never be considered completely safe because to do so would make working with extreme heat second nature, which it surely is not. If you forget for a moment that you are working with a flame that can be as hot as 6,000°F, or that you are around pressurized containers that contain explosive gases, you are flirting with mishap or disaster to yourself and others. While it is impossible to eliminate all the dangers of working with a torch and hot metal, it is possible to reduce the inherent risks by developing safe working habits and religiously following accepted and recommended procedures for handling the equipment.

The purpose of this chapter, then, is to make you aware of the dangers of working with welding equipment, specifically the materials and tools for oxyacetylene welding. With this information in the back of your mind, you should be able to perform all welding tasks with the distinct advantage of knowing how things in general can be accomplished

FIGURE 14.1

Since a welder is always close to the work, it is important that protective clothing be worn.

safely. By knowing and following these guidelines, you will be able to enjoy metal joining and protect yourself from harm.

Clothing

Since you cannot control the direction that sparks fly when welding or cutting, you must protect your skin from possible damage. The best way to do this is to wear specially made clothing that will not support combustion. This type of clothing is available, but it is also quite expensive. Even so, if you plan to do a good deal of welding, it may be to your advantage to invest in fireproof clothing. It is often a requirement for industrial welding (Fig. 14.1).

Cowhide Garments

As you may have guessed, there are alternatives for the home welder in the clothing department. Probably the best choice are garments made of thick cowhide (Fig. 14.2). Currently available are jackets with long sleeves and a high collar that effectively protect the top half of your body from burns. Leather pants or chaps are also a good investment if you plan on doing a lot of welding. While these leather welding garments amount to a sizable investment—about $100 to $150 per suit—this combination is a very good way of protecting your body from the hazards of burns caused by sparks and flying hot slag.

Cotton and Wool Apparel

Cotton and wool clothing are undoubtedly the most popular types for both the casual and professional welder (Fig. 14.3). The ever trusty blue jeans are heavy enough to shed sparks and will not burst into flames if subjected to an open flame. They will smolder, however, but this does not seem to be a problem since the area can be quickly extinguished with a brush of a gloved hand or a few drops of water.

It is important that all clothing worn during welding be free of any trace of oil or grease. These materials burn violently in the presence of pure oxygen.

Wool, in the form of heavy long-sleeved shirts, is quite popular with welders as it is nonflammable, lightweight, and has a relatively long life. Both cotton and wool clothing can be washed frequently without appreciable loss of the material's life. Regular and periodic washing of welding clothing should be done not as much for sanitary reasons as for safety reasons. While the material itself may not burn, some forms of dirt or grime will and should be removed as often as necessary.

Flameproof Clothing

You can flameproof your welding clothing quite simply by soaking in a special fire-retardant solution. First mix up a batch of 12 ounces sodium stannate to 1 gallon of warm water. Then submerge the clothing in this mixture and let it soak for about 15

FIGURE 14.2

Thick cowhide welding jacket and gloves.

minutes. Next, wring out the material, and dip it into a solution of 4 ounces of ammonium sulfate to 1 gallon of water. After about 15 minutes, remove the garment from the solution, wring it thoroughly, and let it dry. Clothing treated in this manner will be flameproof and quite safe to wear when welding or cutting.

Special care is required for flameproof clothing. Only wear the clothing for welding and not for working on the car or truck. When not in use, hang the clothing in an area where it will remain clean. If the clothing ever becomes dirty, as it surely will, do not wash it as this will remove the fireproof impregnation. Instead, have the clothing dry cleaned. When fireproof clothing is treated in this manner, it should last for about six cleanings.

Whenever you are wearing clothing that is not specifically designed and made for welding, you should know that cuffs on pant

FIGURE 14.3

A hat, wool shirt, and heavy pants are suitable for most types of oxyacetylene welding.

legs and pockets on both trousers and shirt are potentially dangerous. These parts of clothing have a tendency to catch and hold hot sparks. It is therefore a very good idea to roll down cuffs on pants and remove pockets on your welding shirt or, at the very least, button shirt pockets closed.

Head Covering

Some type of head covering should be worn by the welder as well as those types of clothing already mentioned. A baseball cap or similar type will not only keep your hair clean but will serve as a protection against hot sparks as well. Since most welding and cutting is done at less than an arm's length away, there is always the possibility of being struck by sparks. While there is little the welder can do to prevent this, he or she can and should take precautions against the hazards of red-hot sparks.

A baseball cap is most effective if worn with the bill on the back of the neck rather than in the conventional manner. This will offer added protection against a hot spark falling down the neck. Additionally, with the bill to the rear, it is easier to manipulate welding goggles or a face shield (Fig. 14.4).

Boots and Shoes

When thinking about clothing for the welder, it is also important to consider what is worn on the feet. Heavy-duty leather boots are the best choice with those without laces at the top of the list. There are any number of boots and shoes designed for use during the welding process. Some styles have the added feature of steel toe inserts that prevent damage to the foot from falling heavy objects.

Lightweight shoes or, heaven forbid, sneakers or tennis shoes should never be worn when welding for obvious reasons. As mentioned, boots are the best protection, but heavy work shoes can also be worn providing that the pant legs cover the tops of the shoes as well. Since you will, in most cases, be concentrating your energy on the area being welded, you may not notice a hot spark burning a hole in the top of your shoe until the damage is done and your foot gets burned. This is another very good reason for wearing heavy-duty footwear.

Eye Protection

Surely one of the cardinal rules of all welding is "Never look at a welding flame or arc unless you have suitable protection covering your eyes." Eye protection is not only necessary to prevent damage to the eye from the injurious rays created during welding or cutting but it is also necessary to protect the eyes from flying sparks and hot slag.

Goggles

There is a wide selection of good eye protection available to anyone planning to weld. In fact, most welding outfits will come with a pair of special goggles in the set. Goggles are available in both high-impact plastic and safety glass. Given the choice, most professionals will choose the safety glass. These goggles do not scratch and tend to offer an undistorted view of the work. Any good-quality pair of safety goggles has lenses that can be removed and replaced if damaged or scratched (Figs. 14.5 and 14.6).

FIGURE 14.4

Tinted goggles and a baseball cap are a good combination for head protection during welding.

Face Shields

Also available are face shields in both clear and tinted styles. Face shields have a few advantages over goggles that are worth mentioning.

FIGURE 14.5

Light-duty welding goggles.

FIGURE 14.6

Good-quality welding goggles will have replacement tinted glass lenses.

The field of view is greater when wearing a face shield, and this will enable the welder to move about the work area more naturally. While wearing goggles, however, the welder may feel quite comfortable while working on a project in a fixed location but will find moving around the workshop difficult. Generally speaking, movement with welding goggles gives the distinct feeling of tunnel vision.

A face shield offers more protection than a pair of goggles simply because the entire face and part of the head are covered (Fig. 14.7). Every face shield I have ever seen could be easily lifted up and away from the face, offering the chance for quick, normal vision if the need should arise. A pair of welding goggles, on the other hand, may not be quite as easy to remove quickly, especially when a hat or helmet is worn along with them.

One of the disadvantages of a face shield is that they are all made from a plastic material, very much the same as a face shield for a motorcycle helmet. Plastic can scratch easily, and surely will over time. Therefore, the welder's field of vision may become distorted somewhat. The only practical solution when a face shield becomes scratched is to replace the lens with a new one.

Filtering Lens

It is a fallacy to think that any pair of tinted glasses or goggles can be used for all types of welding or cutting. A special lens is needed to filter out harmful rays when welding. This filter must be of a type that is suitable for the particular welding or cutting task at hand. A filter's job is to absorb ultraviolet and infrared rays as well as most of the visible light rays. It is important, then, that you use only goggles or other eye protection that has been specifically designed and approved for welding or cutting.

Clear Glass

In addition to a pair of tinted welding goggles, the home welder should have some type of clear eye protection. In most cases, a clear face shield will be more than adequate. Use clear eye protection gear when grinding metal to prevent eye damage (Fig. 14.8). Never rely solely on the shields that are common on most good-quality bench grinders as they will do little, if anything, to protect your eyes from flying sparks and metal particles.

It is a sound principle to keep your clear eye protection gear right next to your bench grinder or other abrasive machinery. Then get into the habit of putting on your clear eye protective gear before turning on the machine. This small act will go a long way in protecting your eyes from damage.

FIGURE 14.7

A tinted face shield designed for oxyacetylene welding and cutting.

QUICK»»TIP

If you wear corrective lenses, speak to your optometrist about buying a special pair of welding goggles tailored to your particular vision requirements. Also available are tinted welding goggles that fit over conventional eyeglasses.

Respirators

Occasionally a welder may come in contact with fumes (both nontoxic and toxic) as well as normal air that is heavy in particulate matter. Probably the most common time for this contact is when grinding metal on a bench grinder, wire brushing metal, or working with a disc sander. While in most cases this "heavy air" will do little harm to the welder, some measures should be taken to protect the respiratory system.

There are a number of different types of respirators available. The range goes from a bandanna secured over the nose and mouth to self-contained breathing masks that would be quite effective during a

FIGURE 14.8

A clear face shield should be worn when working on a bench grinder.

FIGURE 14.9

A respirator is a good precaution when sanding or wire brushing rusted metal.

poisonous gas attack. A little common sense should come into play here. For example, if you are grinding metal that has a lot of surface rust and you see that you will be working on the piece for a long period of time—more than a few minutes—you should wear an adequate covering over your mouth and nose.

One of the more popular means of protecting the respiratory system is an inexpensive cloth mask (Fig. 14.9). You can pick one of these masks up at most welding supply houses. They are also commonly sold in paint stores and lumber yards. Since their cost is minimal, you should pick up a supply and keep them handy around your welding and workshop. Get into the habit of using a respirator whenever you must work in an area that has a heavy concentration of particulate matter in the air. This will reduce the chances of serious respiratory damage as a result of breathing this type of air.

Gloves

Another cardinal rule of welding is "Never pick up a welding blowpipe or cutting torch without a pair of heavy-duty gloves on your hands." By following this simple rule, you will dramatically reduce the chances of burning your hands.

Since at least one of your hands will always be within 1 foot of a flame with a temperature of up to 6,000°F, you should always protect your hands from sparks and reflected heat. The best way to do this is to wear a heavy-duty glove on each hand (Fig. 14.10).

FIGURE 14.10
A welder's hands are always close to the heat, so a good pair of gloves is very important.

There are easily hundreds of different types and styles of gloves designed for welding. Add to this amount a vast selection of general-purpose, heavy-duty work gloves, and you will have little problem in finding a pair of gloves that will effectively protect you while welding or cutting. It is important to keep in mind, however, that not just any pair of gloves is suitable for working around hot metal.

The gloves you wear for welding can be a bit lighter than those you wear for cutting. But, in all cases, the gloves should be heavy and practical. There is a cutoff point. A pair of gloves that are more than adequate for working with a torch may also be too heavy for effectively controlling the flow of oxygen or acetylene. In other words, the gloves may not allow you to turn the control knobs on the blowpipe handle. Consider this carefully when shopping for a good pair of gloves. Purchase a pair that will offer both adequate protection from the heat and will allow you to adjust equipment controls properly with your fingers.

All experts agree that a pair of welding gloves should be flameproof and of the gauntlet type (Fig. 14.11). You will have no problem finding these longer gloves at any welding supply house. At the present time, welding gloves are made from heavy-duty leather, fireproof material, or a combination of the two.

FIGURE 14.11

Gauntlet gloves offer the most protection against sparks and hot slag.

I have found that a pair of welding gloves are easier to work in if they are made from leather. Leather, because it is a natural material, tends to facilitate hand and finger movement a bit better than artificial fibers. Many styles and thicknesses of welding gloves are made from leather. Generally speaking, leather welding gloves do not insulate as well as fireproof gloves. Therefore, the latter may be a better choice when working for a long period of time around molten metal. Keep in mind, however, that fireproof gloves generally tend to be stiffer than leather gloves. There will be a loss of precise finger movement as a result. As I mentioned earlier, there always seems to be some type of tradeoff between gloves that will totally protect your hands and gloves that facilitate finger movement. In the end, the choice must be yours. You will probably end up having at least one pair of leather gloves and one pair of fireproof gloves.

Aprons

One last piece of clothing that is worth mentioning is an apron for welding. A good-quality apron, designed specifically for use during welding, offers one very good alternative to traditional heavy-duty clothing. Aprons are currently available in many different styles, materials, and price ranges. To be effective, the apron should protect as much of the welder's front as possible while enabling the welder to have free movement of arms and body.

Leather and fireproof fabric are probably the two most common materials used to make welding aprons. Both materials are quite effective at reflecting heat and deflecting hot sparks while welding or cutting. Look at the aprons offered at your local welding supply house. You quite possibly will find one that is suitable for all your welding protection needs. Obviously all of the care and use requirements of clothing mentioned earlier apply to aprons designed for welding (Fig. 14.12).

Fires and Fire Equipment

Because the possibility of a fire always exists for the welder, he or she should have some means of putting a fire out if one should start. Before you can put out a fire, you must know a few things about fires in general. The old military saying "Know your enemy" is one of the most effective ways of dealing with fires. As a way of making you more aware of fires in general, we need to discuss the basic components of fires.

FIGURE 14.12

A bib apron provides adequate protection of clothing.

Fuel, Oxygen, and Heat

Before a fire can take place, three things are needed: fuel, oxygen, and heat. If any of these three components are removed, a fire cannot continue to burn. Whenever a fire breaks out, you must make a quick decision as to which of the three parts of that fire can be eliminated most easily, thus extinguishing the fire. The three obvious means of doing this are removing the fuel (which can often be near impossible), cooling the material that is burning, or removing the oxygen by smothering.

All fires have two things in common, heat and oxygen, and these can be an aid to you in fighting any fire. The third component of a fire, fuel, can vary from one fire to another. It is because of the possible variation of fuel in any given fire that all fires are grouped into four possible types: Class A, B, C, or D. You should know about each class of fire in order to determine the best way to extinguish any particular fire.

CLASS A FIRES

Class A fires are those in which wood, paper, clothing, and similar materials are burning. Generally speaking, water is the best and most effective way to fight a Class A fire. Water will penetrate into the burning material and cool it. Water will also, at least in part, seal the material from the oxygen in the atmosphere, although to a much lesser extent than other firefighting materials.

CLASS B FIRES

Class B fires consist of flammable liquids such as oil, gasoline, paint, and similar materials. As a rule, once the flame is extinguished in a Class B fire, the fire will go out. You must, in effect, remove the oxygen or separate the burning material from oxygen to extinguish a Class B fire. Only the vapor on top of the liquid surface is burning and not the material itself in a Class B fire. For example, diesel fuel burns on the surface as the vapors evaporate into the surrounding atmosphere. The fuel itself does not contain oxygen. Separate these fumes from the oxygen in the atmosphere and the fire will go out.

The most effective means of fighting a Class B fire is, of course, to smother the fire, thereby separating the vapor, which is burning, from the fuel or burning material, which is not burning. Foam, powder, sand, and other types of nonflammable materials are most effective because of the heavy smothering action. Once the burning material is separated from the oxygen in the atmosphere, the fire will go out almost immediately.

In no case should water ever be used to try and extinguish a Class B fire. Most flammable liquids will float on top of water and continue to burn. Water will not, therefore, extinguish a Class B fire. In fact, when water is thrown on a Class B fire, it may actually float the flames and fuel over a large area, which could dramatically spread the fire and prevent putting it out.

CLASS C FIRES

Class C fires are electrical in nature. This class of fires includes electrical wiring, motors, switches, machinery, and appliances. A smothering action is the best means of fighting a Class C fire after the power has been turned off. For our purposes, that means a nonconductive material, generally a special type of fire extinguisher. Because of the possibility of electrical shock, water should never be used in an attempt to extinguish an electrical fire.

CLASS D FIRES

Class D fires are rather rare and almost nonexistent for the home welder. These fires are caused by magnesium shavings. The only way to put out a Class D fire is to use a specially designed fire extinguisher specifically labeled for Class D fires. If the home welder ever plans to work with combustible metals, he or she should have a fire extinguisher of this type handy.

Any fire will fall into at least one of the above classes of fires: A, B, C, or D. Probably the best way to fight and extinguish a given fire is to use a fire extinguisher that has been designed to put out a specific class of fire, or all classes of fires.

Fire Extinguishers

As you may have guessed, there are fire extinguishers for each class of fire. There are also fire extinguishers that are suitable for more than one class of fire. These will be clearly labeled. Probably the best choice for the home welder is a type generally referred to as *multipurpose* or *all-purpose, fire extinguisher.* These are clearly labeled for "ABC Fires" (Fig. 14.13).

The best types of fire extinguishers will have a pressure gauge on top of the unit that will indicate the condition of that particular fire extinguisher. Some are more elaborate than others, but all that is really required is a gauge that will indicate whether or not the unit will work when needed. A gauge like this will give you valuable information at a glance, *recharge* or *operable,* for example (Fig. 14.14).

Even though your fire extinguisher may have a gauge, you should have it checked by a qualified repairperson once a year. This will ensure that the unit is, in fact, in good condition and that it will be an effective means of fighting a fire.

If you ever use a fire extinguisher, you should have it refilled, even if you use it for only a few seconds. What would be the point of having a fire extinguisher around your shop if it were only half full or not properly charged? Your property and possibly you life may depend on your ability

FIGURE 14.13

This fire extinguisher is designed to fight Class A, B, and C fires.

FIGURE 14.14

A fire extinguisher should have a gauge that tells if the unit is operable or not.

FIGURE 14.15

Sand and water can be used to fight certain types of fires.

to quickly extinguish a fire. This is reason enough to make certain your fire extinguisher is in first-class working order at all times.

Sand and Water

In addition to a good-quality and multi-purpose fire extinguisher, you should also have a bucket of dry sand and a pail of water in your workshop or in the area where you are working. Sand can be used quite often to extinguish a Class B fire because it will quickly smother the flames. A bucket of water should only be used for putting out a Class A fire for those reasons previously mentioned. Nevertheless, both sand and water can be used to quickly extinguish certain types of fires and offer an alternative to your fire extinguisher (Fig. 14.15).

Obviously, your best defense against fires is to prevent them from happening. That means working only in safe surroundings. Since the next chapter covers many aspects of the home welding shop, you will have to turn there for more information on the best type of surroundings in which to weld or cut.

Burns

Since we are discussing fires, it will be worthwhile to mention a few things about burns. To say the least, burns resulting from welding or cutting are not a pleasant topic for discussion. Nevertheless, the home welder should know what to do in the event of a serious burn.

Generally speaking, all burns can be grouped into three rather large categories. Since treatment of any one particular burn will depend on the nature of that burn, we will discuss all types of burns. All burns fall into either first-, second-, or third-degree burn categories.

First-Degree Burns

First-degree burns result in red skin and possibly a small amount of pain. Sunburn is usually a first-degree burn as is red skin, due to reflected heat from welding or cutting. Generally speaking, first-degree burns are more uncomfortable than serious. The best treatment for first-degree burns is to relieve the slight pain by applying a burn ointment to the area. If the burn covers a large area, see a doctor as soon as possible. Generally, the pain will subside in a few hours. If pain persists, seek medical help.

Second-Degree Burns

Second-degree burns are a bit more serious and are characterized by blisters. Treatment for second-degree burns should *not* include covering the blisters with a burn ointment. You should prevent the blisters from breaking by covering the area with sterile gauze and holding it in place with a bandage. Second-degree burns are serious, and medical attention should be given by qualified personnel as soon as time permits. The greater the area of the second-degree burn, the more potentially serious the problem.

Third-Degree Burns

Third-degree burns are the most serious of all burns and are characterized by burned or charred flesh. Because of the nature of welding and cutting equipment, third-degree burns are common. It will therefore be to your advantage to know what to do if this type of burn should happen to you or someone else in the welding shop. Know in advance that third-degree burns which cover more than 50 percent of the body often result in a fatality. Therefore, quick action on your part may make a difference.

There are a few cardinal rules that should be followed when dealing with third-degree burns:

- Do not move the person unless absolutely necessary.

- Do not apply burn creams or salves to cover the burn.

- Do not remove burned clothing. Leave it in place to be removed by qualified medical personnel.

- Treat for shock by covering the victim with a blanket or other suitable material. Lay the person down with feet elevated, if practical.

■ Seek medical attention as quickly as possible, either by rushing the victim to the hospital or calling in emergency medical assistance.

All burns require attention as they are all potentially serious. It is important to keep in mind that quick action on your part may make a difference, so act accordingly.

Handling Welding Equipment

The equipment used in welding and cutting is potentially dangerous and should always be treated with the respect that it is due. Failure to do so may result in mishap. Many points for safe operation can be found throughout this book. The majority of the remainder of this chapter will summarize those practices that should be standard operating procedures for anyone working around or with welding equipment.

Oxygen and acetylene cylinders, as well as other fuel gas containers, have properties that can cause serious accidents, injuries, and even death if proper precautions and safety practices are not followed. It is therefore very important that you become aware of these safety precautions. Make them a part of your welding habits.

Because both oxygen and acetylene are in pressurized containers, they must be handled with respect. It is very important that these cylinders be securely held in an upright position at all times. Probably the best way of doing this is to chain both cylinders to a special hand truck designed to hold oxygen and fuel gas cylinders. These hand carts are available where you purchase your oxygen and acetylene (Fig. 14.16).

FIGURE 14.16

Oxygen and acetylene cylinders should always be securely fastened so that there is no chance of falling. A wheeled hand cart is probably the best means of storage for your tanks.

If the tanks are to remain in one location, make certain that both are secure by chaining them to an immovable object like an interior wall or column. Never allow either type of cylinder to stand by itself for any period of time. This will be your best insurance against accidental tipping of the cylinder.

Keep all organic materials as well as flammable substances away from both oxygen and acetylene cylinders. Remember that even normal soot and dirt can constitute a combustion hazard around these gases.

Never use either oxygen or acetylene (or other fuel gases) without first attaching an approved regulator to the tank. A full tank of oxygen will have an internal pressure of about 2,200 pounds. Full acetylene cylinders will have an internal pressure of up to 400 pounds. It is therefore extremely important that only a suitable regulator be used to reduce these pressures to recommended working pressures in the lines (Fig. 14.17).

FIGURE 14.17

Always use the proper regulator for oxygen and fuel gas.

Handling Oxygen

In addition to the general rules for safe handling of both oxygen and acetylene, there are other recommendations that apply specifically to oxygen. Never allow oxygen to come in contact with oil, grease, kerosene, paint, tar, coal dust, or dirt. Oxygen supports combustion and can burn violently in the presence of these and other flammable substances. You should never handle oxygen or equipment that may come in contact with oxygen—such as the oxygen regulator—with dirty hands, grease-stained gloves, soiled clothing, or soiled rags.

Before attaching a regulator to the oxygen cylinder valve, wipe the fitting and threads with a clean, dry cloth (Fig. 14.18). You should also crack the valve on the oxygen cylinder to blow out any dirt or other foreign matter that may have found its way into the valve. When cracking the oxygen cylinder valve, do so quickly. Position your body so that the oxygen in the tank will not spray on you or your clothing. Stand to the side of the tank valve (Fig. 14.19).

Never use oxygen for ventilation or as a substitute for compressed air. You should never blow dust off anything with oxygen. Cleaning your clothes off with a blast of oxygen and then lighting up a smoke is one of the quickest ways to end up in the hospital.

Always refer to oxygen as "oxygen" and never as "air." This will avoid the possibility of confusing oxygen used in the welding process with compressed air.

Never use oxygen for other than its intended purpose. Oxygen that is designed to be used for medical purposes such as therapy for respiratory ailments will have U.S.P. stamped on the cylinder. All other types of oxygen are for industrial uses only and should not be used for oxygen therapy.

When transporting oxygen, do so only with the container in an upright position in an open vehicle. Never carry an oxygen cylinder laying on its side in the closed trunk of an automobile.

FIGURE 14.18

Always wipe the threads on regulator fittings with a clean dry cloth before attaching to the fuel or oxygen cylinder.

Handling Acetylene

Acetylene cylinders are equally dangerous. All of the guidelines for safe operation and use of oxygen cylinders should be followed for this fuel gas as well. In addition, you may recall that all acetylene cylinders have a safety plug in either the top or bottom of the cylinder. This safety plug is designed to melt at a temperature of about 212°F. It is a safety feature intended to reduce the hazards of an explosion in the event of a fire (Fig. 14.20). Because this safety plug has such a low melting point, you should never work with an open flame around a cylinder of acetylene. For the same reason, you should never pour boiling water on top of an acetylene valve in an effort to thaw out a frozen valve. This can happen when an acetylene cylinder has been stored in an unheated room or outdoors during cold weather. At all times you should protect the fusible safety plug in acetylene cylinders from damage. In addition, store, transport, and use acetylene in a vertical position.

FIGURE 14.19

Stand to the side when you crack an oxygen or acetylene cylinder valve.

Review of Equipment Setup

When setting up welding equipment, specifically oxygen and acetylene cylinders, follow those guidelines set down in Chapter 1. A quick review of those steps may further drive home the importance of proper handling of all welding equipment.

Oxygen Connections

Make certain that the fuel and oxygen cylinders are securely held in one place that will permit them to have a safe, happy life. Crack the

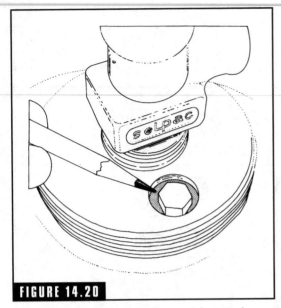

FIGURE 14.20

The safety plug in the top of an acetylene tank.

valve of the oxygen cylinder before attaching the special oxygen regulator. Wipe the threads of the fittings of both the regulator and cylinder valve before fastening the regulator in place.

Tighten the regulator with a wrench designed for the purpose (Fig. 14.21). Undertightening will result in a leak, and overtightening may cause the brass fitting to become strained and weakened.

Connect one end of the oxygen hose (green in color) to the regulator and the other end to the oxygen fitting on the torch handle. Snug up all connections with a suitable size wrench.

Check to make sure the oxygen control knob on the blowpipe handle is closed before opening the valve on top of the oxygen tank. In addition, open the regulator control lever until there is no resistance in turning of the lever. The diaphragm is now open. Open the oxygen control valve on the cylinder slowly (Fig. 14.22). This will ensure that the delicate oxygen regulator does not receive a surge of pressure from the cylinder. Remember that a full tank of oxygen will contain approximately 2,200 psi of internal pressure. Once a bit of oxygen is in the regulator and the high-pressure gauge registers the pressure in the tank, open the valve two turns. At this time, give a quick check to the regulator to make certain that it is not leaking. If you suspect it is, use the soapy water test.

In most cases a leak is a result of not tightening a connection properly. If you discover a leak, first turn off the valve on top of the tank. Then try snugging up the connection with the proper size wrench. Put pressure into the system and check the suspected area (Fig. 14.23). If the leak persists, there may be internal damage to the regulator or fitting. It should therefore not be used until the problem has been sufficiently corrected.

Acetylene Connections

The procedure is essentially the same for hooking up acetylene connections. Keep in mind, however, that all connections for acetylene are of the left-handed type and will be marked with a groove cut around the circumference of the brass fitting (Fig. 14.24).

After the recommended working pressures have been set on the regulator, open the torch handle momentarily to see that acetylene is, in fact, flowing through the line. You should also keep one eye on the regulator dials and make sure that they move slightly, if at all.

Working with Oxyacetylene Equipment

When working with oxyacetylene equipment, there are a few points to keep in mind. The first is that you should turn off the flow of both oxygen and acetylene at the cylinders if you stop work for more than a few minutes. The second is that if you are shutting down the equipment for more than a half hour, you should, in addition to turning off the flow of oxygen or fuel gas, also bleed the lines. Clear the acetylene line first by opening the control knob on the blowpipe handle. Watch the acetylene regulator when you do this. In a matter of moments both gauges should register zero. Then turn off the acetylene control knob on the blowpipe handle. After you have cleared the line of acetylene, follow the same procedure for cleaning the oxygen line (Fig. 14.25).

Obviously, when you are shutting down the oxyacetylene equipment, you should make certain that there is no open flame in the area. When working with oxyacetylene equipment, it is important to

FIGURE 14.21

Tighten the regulator fitting with a suitable size wrench.

FIGURE 14.22

Always open the cylinder valve slowly so that the regulator will not be damaged by a sudden surge of pressure.

FIGURE 14.23

Set the working pressure by turning the lever until the desired pressure is reached.

FIGURE 14.24

All fuel gas fittings have left-handed threads and a notch around the fitting nut.

FIGURE 14.25

You can clear the oxygen line by depressing the oxygen cutting lever on the torch. The supply of oxygen must be shut off at the tank first.

always have a clear path to the tanks of oxygen and fuel gas. In the event of a mishap, one of the first things you should do is to turn off both oxygen and acetylene at the top of both tanks.

Another good practice when welding is to always make sure that the hoses carrying oxygen and acetylene do not lie under the work. A hot spark or piece of hot slag falling on a hose could easily burn its way through the hose and result in a fire.

Store your welding equipment and tanks in a safe place. This equipment, for example, should never be left within easy reach of children. Both oxygen and acetylene cylinders come with a special valve protection cover. This cover should be in place over the valve whenever the regulator is not attached. If there is any possibility that children will be in the area where oxygen or acetylene is stored, the valve protection cap should be made secure and, at the very least, tighter than small hands can loosen (Fig. 14.26).

Hose, torch, cutting attachment, and tips should all be stored in a safe place as well. A cabinet that will keep this equipment dust free is best. It will go a long way in preventing costly repair bills as well as mishaps around the workshop.

Welding Hazards

In summing up this chapter on safety, I think it will be helpful to briefly outline the hazards that are always present in welding. As a welder, you should know these hazards and continually strive to take them into consideration. This will be your best insurance against an unfortunate and possibly tragic experience.

■ **Fire.** The possibility of fire always exists. You should therefore always have at least one means of extinguishing a fire if one should break out.

■ **Burns.** Because of the nature of welding equipment, a serious burn can happen at any time. Always wear protective clothing to reduce the chances of burns to yourself or others in the shop.

■ **Eye injury.** Always wear tinted eye protection when welding, cutting, or even simply adjusting the flame of your blowpipe. Additionally, you should wear clear eye protection when performing welding-related tasks such as grinding metal surfaces (Fig. 14.27).

FIGURE 14.26

The safety protection cap should always be on a cylinder of oxygen or acetylene whenever the regulator is not attached.

■ **Explosions.** Always handle, transport, and store oxygen and fuel gas cylinders with the respect due them. Check all connections to make certain that no fuel gas or oxygen is leaking into the atmosphere. Exercise care around other containers with flammable materials.

■ **Harmful or poisonous gases.** Welding or cutting some types of metals, such as galvanized steel, may result in a harmful gas. Make certain that the welding area is adequately ventilated at all times. Wear a respirator over your mouth and nose whenever grinding or welding potentially dangerous metals.

■ **Mishaps.** A danger always exists when working in unsafe surroundings or as a result of unsafe work habits.

FIGURE 14.27

Eye and head protection should always be worn when welding or cutting.

- **Failure to exercise caution and common sense.** Think before you act and you will eliminate 95 percent of the danger associated with welding and cutting operations.

A wise welder knows and practices safe work habits. He or she is constantly aware of the dangers involved in welding and does everything possible to make welding and cutting a safe and enjoyable means of accomplishing various metal joining tasks.

The Home Workshop

In addition to oxyacetylene welding equipment—blowpipe, cutting attachment, safety equipment, hoses, tanks of fuel gas and oxygen, and related accessories—the home welder requires a safe space to work. If you think you can take care of all of your welding projects in the back of your garage or in your basement, you are flirting with danger. Welding, because of the inherent risks, requires that the work always be done in a special area. After all, we are dealing with controlled fire and very hot metal when welding or cutting. If special precautions are not taken, unfortunate accidents may result.

Outdoor Workshop Advantages

When just beginning, chances are very good that you will be practicing cutting and welding outdoors. This makes sense on a number of levels. Even though there are some drawbacks to working outside, you should find an outdoor workshop more than adequate for your practicing needs.

Ventilation

Probably the greatest advantage of working outdoors is that there is always plenty of ventilation. Even on windless days, smoke and fumes from welding and cutting will quickly drift up and away from the area.

As you may already have discovered, the resultant smoke from welding, even though nontoxic, can be upsetting to the system. It is therefore a good practice to inhale as little smoke or fumes as possible.

Equipment Setup

Setting up equipment, as well as shutting down the equipment, can be safely done outdoors as well. For example, when you crack the oxygen or acetylene valves prior to attaching the regulators, these gases will quickly disperse into the atmosphere and pose little danger. The same "clearing of the air" also takes place quickly when bleeding the oxygen and acetylene lines while the equipment is shut down. These gases will simply drift away. When you are working indoors—in a workshop, for example—you must make certain that these gases are exhausted from the area by some artificial means such as a ventilator fan. Later in this chapter we will discuss the various means of introducing fresh air, and plenty of it, into an indoor welding workshop.

Minimal Cost

Another advantage of an outdoor workshop for welding is that the cost of setting up such a shop will be, in most cases, quite minimal. In addition to the necessary oxyacetylene welding and cutting equipment, you will also require a worktable. Since a worktable is one of the basic components of any welding shop as well as a real aid to working with oxyacetylene equipment, you will want to build one as soon as you have mastered some of the basic cutting and welding techniques discussed earlier in this book.

Welding Table

To build a simple welding table, you will need 20 feet of 2- by 2-inch angle iron, 20 feet of 1-inch-wide steel strap, and 24 firebricks 8 by 4 by 2 inches (see Table 15-1).

The dimensions for a standard welding table are given in Fig. 15.1. Begin by cutting four pieces of angle iron, each 28 inches long. Then cut the remainder of the angle iron into two 24-inch lengths and two 32-inch-long pieces. The angle iron (actually angle steel) makes up both the legs and the frame for the top of this welding table. You must cut precisely for best results.

**Table 15-1 Materials list
for the welding table.**

Quantity		Size (inches)	
4	28	2"-×-2" angle iron
2	24	2"-×-2" angle iron
2	32	2"-×-2" angle iron
4	24	1" steel strap
5	32	1" steel strap
24		8"-×-4"-×-2" firebricks

FIGURE 15.1

Dimensions of the welding table.

Assembly

Begin assembling the table by laying out the top pieces and welding them together (two 24-inch- and two 32-inch-long pieces). You will be welding a rectangular frame that when finished will measure 24 inches wide by 32 inches long. Do not overlap the pieces but, instead, use the simple butt joint. You may find it easier to first tack weld the pieces together before welding. Work carefully to ensure that the four corners of the table remain square.

After the top of the welding table has been assembled, attach the 28-inch-long legs. Attach one leg to each corner, and tack weld on the outside of the frame. When completed, you will have the basic frame of the table. All that remains is to weld the steel strap—four 24-inch-long pieces and five 32-inch-long pieces—inside the frame top. Placement is important because the steel strap provides support for the firebricks that will be the working surface of this table.

It is probably easiest to begin by welding the 24-inch-long steel strap first. Space the strap at 8-inch intervals, and tack weld them in place. Next, the 32-inch-long strap is either laid on top of the cross members or woven through them for additional strength. Just the ends of these longer strips are tack welded until all straps are in place; then you can go back and tack weld all joints where straps overlap. You must place the longer straps at 4-inch intervals so that the edges of the firebrick will lie on them. If you are in doubt as to the placement of the steel strap, simply lay a few firebricks into the frame to give you a clear picture of the exact layout of the straps (Fig. 15.2).

As you can see from Fig. 15.1, the worktable will have a finished height of 28 inches. If you are taller or shorter than average, you may want to build a table of a different work height. To do this, simply adjust the length of the legs. It may be of interest to you to know that you will have some extra 2-inch angle iron if you buy the listed 20 feet. Actually you will have an excess of about 16 inches, so you will have little problem increasing the basic height of this table by about 3 inches with the

FIGURE 15.2

Tack weld the steel strap around the inside of the table frame. Proper spacing is very important.

materials you have on hand. Twenty-eight inches is, by the way, the standard height for a work surface, and most people find this height about the best overall working height.

Modifications

Modifications are, of course, possible with the basic plans offered above. Some additions you may want to include are roller wheels, which can be locked, for each of the four legs. The addition of wheels will make the table that much more versatile as it can be moved about the workshop quite easily.

You may also want to add additional angle iron midway between the top of the table and the floor. Then you can add a shelf to hold scrap metal or other materials. Keep in mind, however, the potential fire danger of storing materials under a table that will be used for cutting and welding.

Another addition might be a ¼-inch-thick steel plate, 24 inches long and 8 to 12 inches wide. Attach this plate to one end of the worktable, and you will increase its versatility by providing a solid and unbreakable work surface.

You can also attach hooks on one end of the tabletop edge. These can be used to hang various and often-used tools such as a hammer, clamps, and spark lighter. The point is to make your worktable useful to you as a welder. I would suggest that you build the basic table and use it for a period of time (Fig. 15.3). Then, after you have a clear picture of your work surface requirements, add personal touches to make your work easier and possibly more enjoyable.

Outdoor Workshop Sites

Your outdoor workshop can be almost anywhere around your home that space permits. As a rule, you must have access to electricity for running a grinding wheel, drill, and possibly other equipment. You will also find it much easier if you can set up your equipment and begin work quickly rather than making a half-dozen trips to the house or garage to move the

FIGURE 15.3

The finished welding table made from angle steel and steel strap.

required materials and tools out to the work area. In most cases, this will mean that your outdoor work space will be close to your house or garage. It will be helpful to discuss some of the possible sites for an outdoor workshop while pointing out some of the requirements of a good spot as well.

Carport

A *carport* can make an ideal outdoor workshop. You will be able to leave your welding equipment and tanks in one spot without having to worry about the elements of wind and water playing havoc with them. You will also be able to store scrap and unused metal out of the weather. Ventilation will be of little problem as most carports are open at least on two sides. You must make certain, however, that flammable materials are not in the area. In some cases, this will mean leaving your automobile and lawn mower in another location, as well as paints or other materials, when welding and cutting. If a concrete floor exists in your carport, it should be clean and free from oil or grease stains. If wooden walls exist in the carport, you should locate your welding table as far as possible from them. You can cover existing walls with asbestos board, commonly sold in sheets from 2 to 4 feet wide and from 4 to 8 feet long.

FIGURE 15.4

When oxygen and acetylene tanks are stored outdoors, they must be chained to a solid object and covered when not in use.

Cement or Stone Patio

If you are not fortunate enough to have a covered carport to use as a welding shop, consider a cement or stone patio that is close to the house. Keep in mind, when using an uncovered area for welding projects, that you will be limited by good weather as to when you can work. You must also rig up some means of covering the oxygen and acetylene cylinders if they are to be left outside (Fig. 15.4). In some areas of the country you can get by quite nicely by covering the equipment with a plastic tarp when not in use. In other parts of the country, this may not be the best protection. In the Great Basin (western deserts), rainfall is almost nonexistent but, in the Pacific Northwest inclement weather is quite common, for examples. You can also work on bare ground, if necessary, providing that all materials such as fallen leaves are removed from the area each time you plan on using the outdoor shop.

Outbuilding

If you live on a farm or ranch, you may quite possibly have an outbuilding that can be used for a welding shop. You may also find that a number of welding tasks need to be accomplished in places other than the welding shop. In cases such as these, you will find it very handy to have your oxygen and acetylene tanks on some type of portable hand truck. Then you will be able to take the welding to where it is needed rather than the other way around.

Whenever you work outside, which in many cases will be on bare ground, you should clear the area of any materials that might catch on fire during welding or cutting operations. If you are working during the dry part of the year, it may be a very good idea to wet the earth with a garden hose, if possible, before beginning work. Keep the hose close to the area you are working in so you will be able to quickly extinguish any fire that might occur. Needless to say, other firefighting equipment should be handy as well.

Indoor Workshop Needs

If outdoor space is not available or if you have suitable indoor space for welding, such as a garage, there are a number of things that you should take into consideration before any welding or cutting project is started. As I mentioned earlier, adequate ventilation is an important prerequisite of any welding area. When you are working indoors, you must make certain that the air in the space is fresh and moving. In most cases a common exhaust fan, the type common in most kitchens, will do the job quite nicely (Fig. 15.5). If you are working in a large area, you may find it necessary to install a heavy-duty, large-capacity fan to ventilate the room. The point is that all smoke and fumes must be removed from the work area as quickly as possible (Fig. 15.6). You

FIGURE 15.5

You must install an exhaust fan in an indoor welding shop to remove the fumes from welding and cutting operations.

should also remember that fresh air has to come from the outside of the room and not just be recirculated. You may, in addition to installing an exhaust fan, also have to install a fresh-air vent to introduce outside air and exhaust interior air.

Fire Station

Your home workshop, whether indoors or out, should have adequate firefighting equipment that is accessible and ready to use at all times. In most cases, it is best to designate one particular area as a *fire station.* Here you should keep a fully charged fire extinguisher, bucket of sand, and a pail of clean water. The fire station should never be used for anything like storage of scrap metal. There should always be a clear path to the firefighting equipment. That way, if you are ever in need of firefighting equipment, you will not only know exactly where it is but you will also be able to get to it as quickly as possible. A fire station may quite possibly save you from the severe consequences of an out-of-control fire in your workshop.

Accessibility

Still another requirement of an indoor welding workshop is that you must have accessibility for delivery of supplies such as tanks of fuel gas and oxygen. If your welding shop is in a garage, there should be little problem in getting materials and supplies into the shop. If your welding shop is in your basement, however, you may find deliveries a real exercise in equipment manipulation. Consider easy access as one very important prerequisite of any site you are planning to use as a welding shop.

Lighting

Lighting is important in any workshop, and a welding shop is certainly no exception. It is a fallacy to think that you can work in a welding shop under a single light bulb. In addition to good lighting over your welding table, you must also have direct light over your bench grinder and other work areas, such as where you solder or where your bench vise is located.

Probably the best type of lighting for a welding shop is fluorescent (Fig. 15.7). Two 4-foot-long tubes over the welding table should provide you with plenty of illumination. Fluorescent lighting is also the cheapest type of lighting to operate and run over time. For lighting of specialty areas, such as over the bench grinder, it is usually best to install a lighting fixture over each particular area (Fig. 15.8). Some possibilities include a clamp-on reflector lamp or small fluorescent fixture. If you find that you will require a number of lighting fixtures, it will probably be best to install fluorescent lights as they generally run cooler and, as mentioned, are less expensive to operate in the long run.

While we are on the topic of electrical lighting, it may be a good time to mention that the indoor welding shop should have a number of electrical outlets. Various tasks will be easier to accomplish if you can plug equipment into any one of several electrical outlets. Considering that there are a few electrical tools that might be commonly used around the welding shop, you will want to make provisions for them. You should have a few outlets in addition to those already in use by stationary equipment and machinery.

You may also want to install a 220-volt electrical outlet in the welding shop. Most arc welding units require this higher line voltage for operation.

FIGURE 15.6

One manufacturer's solution to the smoke problem when welding is a smoke exhaust gun system. This system is used in industrial welding.

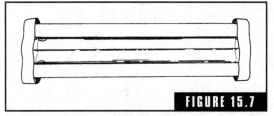

FIGURE 15.7

A fluorescent lighting fixture will provide good general lighting in the work area, at a lower cost than incandescent lighting.

FIGURE 15.8

It is important to provide adequate lighting over your grinding bench. This clamp-on light is inexpensive and does the job.

Storage Space

Another consideration when planning a welding workshop and one that will become more apparent as you spend more time welding is that you will almost always require more space than you have. Such is life. Just how much space you really need will depend on your work habits and even the type of work you plan on doing. In all probability you will be limited to a garage or other outbuilding plus the area outside the shop. Over a period of time you will undoubtedly accumulate metals, both new and scrap, and you will want to store them out of the elements. You will also probably find yourself with a number of tools and machines that are necessary for various metal cleaning or joining tasks.

The obvious solution to "all that junk," as my wife is fond of saying, is to build storage units in one area of the workshop. Sturdy shelves can be used for storing small pieces of metal. Long pieces, such as angle iron or strap, are best stored horizontally and flat on some type of rack system. Plate and sheet metals can easily be stored standing on edge in some type of vertical shelving system. Odd sizes and shapes of metal are usually a problem to store. Often a metal garbage can with three or four caster wheels welded on the bottom will take care of the problem. In all cases it is always best to store metal off the ground, even if a concrete floor exists, to prevent rust on the metal (Fig. 15.9).

Chances are that most of the metal you work with will be steel in one alloy or another. You may also have some nonferrous metals around the shop as well. It will be better, in the long run, to store all similar metals together. Some type of labeling system will be helpful in identifying these metals.

If your workshop is a small one, you will probably be forced to store much of your metal outdoors. To keep the metal from rusting, you must cover it with plastic or a canvas tarpaulin. Keep in mind that the best place to store metal is in some type of closed shed. The more serious you become about working with metals, the closer you will get to building a storage shed for your supply of raw materials and metal.

Accessories

There are any number of useful projects that you can build around your welding shop. In addition to having specific areas for storing metals, you will also need storage cabinets for tools and accessories. Since cabinet building is really beyond the scope of this book, I will not deal with these woodworking projects here.

Mobile Cart

One project that I will briefly discuss, however, is a storage unit for welding and brazing rods. Plans for a basic unit are given in Fig. 15.10. This unit can be made into a mobile cart by simply adding four caster wheels and a bottom panel to the basic unit. You will find a cart quite useful for keeping your rods stored neatly as well as keeping them handy when you need them.

FIGURE 15.9

A storage rack for lengths of steel pipe is easy to build and keeps the materials off the ground.

Anvil

Another useful accessory for your welding shop is some type of *anvil.* An anvil is quite handy for bending or flattening small work with a hammer. You can make one simply by cutting one end of a piece of scrap steel rail in the shape of a traditional anvil. A piece of rail about 16 inches long will be more than adequate for your needs. After the basic shape has been cut out of one end of the rail, pierce cut four holes on the bottom with your cutting torch. Use these holes for fastening the anvil to a solid surface. One very good base for your home

20″

20″

20″

4×20″

4×20″

20″

FIGURE 15.10

A welding and brazing rod holder table.

anvil is a section of tree about 18 to 24 inches in diameter and about 20 inches high. The log will absorb most of the shock when hammering and will also help to deaden the noise as well. Finish off your anvil with a portable grinding wheel or disc sander.

Local Codes

One last thing to consider about your home welding shop are the local codes. In some residential areas, a commercial welding shop is not permitted. It will be in your best interest to check with a suitable governing body to see if any such restrictions exist in your area.

As a side note about restrictions in certain areas, you should know that if you call your shop a "studio" rather than a "welding shop," you may find that the same restrictions do not apply. In many instances your workshop will be a "studio" in the sense that it will be a place where you can express yourself through the art of metalworking.

Summary of Workshop Requirements

Choosing a location for your welding shop is a major consideration. If you limit yourself to small projects and only work occasionally, you will obviously be able to work outdoors. However if you are planning to devote a fair share of your time to welding and cutting, you will need a space that can accommodate your needs. In an effort to help you decide if a particular area is suitable for general welding and cutting needs, consider the following eight requirements of a safe and useful welding workshop:

- *Ventilation.* Some means of exhausting smoke and gases must exist or be installed in the area. This exhaust system must have the capability of quickly clearing the air in the workshop as well as supplying adequate fresh air.

- *Fireproof construction.* The walls, floor, and ceiling of the welding shop must be covered with a noncombustible material to prevent fire.

- *Work area.* A good welding shop will have different areas for specific metalworking tasks. These might include a cutting and welding table, soldering workbench, metal grinding area, and possibly a separate area for finishing or painting metal.

- *Lighting.* Before you can expect to work in a workshop, you must provide adequate lighting.

- *Fire station.* Every welding workshop should have firefighting equipment in a central location.

- *Storage.* Welding, cutting, and related tasks such as grinding can be dusty affairs, to say the least. A good welding shop will have storage facilities not only for metals but for tools and equipment as well.

- *Access.* An effective welding shop will have wide doorways that permit easy movement of equipment and supplies.

- *Cleanliness.* A safe welding workshop is easy to keep clean and should be kept that way at all times. This will do a lot to prevent fire and also maintain pleasant working conditions.

In the final analysis, you will make the decision as to the best location for your welding shop. Obviously, it will be to your advantage to choose a location that is both adequate for your needs and enables you to safely practice the art of welding.

Metal Projects

As you become more proficient at joining metals with the oxyacetylene process, you will find an endless supply of metalworking projects to keep you busy. In this chapter we will look at 10 different metalworking projects that you can build with your welding equipment.

Metal Post Pounder

Metal fence posts are widely used for fencing, especially in the western states where they are used for barbed-wire fencing systems. Metal fence posts are much less expensive than wooden fence posts and last for decades. Unlike wooden fence posts—which require a hole to be dug—metal fence posts can be pounded into the ground with the aid of a post pounder.

You can easily make a metal fence post pounder (Fig. 16.1) in your home workshop in a few hours, and it will last for years. Often you can find the materials for making this handy tool around the workshop.

Begin by cutting the ¾-inch galvanized water pipe into two 24-inch-long pieces and four 12-inch-long pieces. Cut the ends of the 24-inch pieces and one end of the 12-inch pieces at a 45-degree angle. These cuts will enable you to make beveled joints on the handles. Use

Exploded view
showing pieces

FENCE POST POUNDER

FIGURE 16.1

Metal post pounder.

MATERIALS

- 36-inch-long, 2½-inch- (or wider) diameter, water pipe
- 8 feet of ¾-inch galvanized water pipe (8 feet of ⅝-inch re-bar may be substituted)
- ⅜-inch steel plate

a hack saw or other suitable saw for the cutting.

Next attach a 12-inch piece of pipe to each end of the 24-inch piece of pipe. Fit the pieces together and clamp before welding. When complete, this will be one side handle for the post pounder. Repeat for the second handle.

Once the handles have been welded, attach one to each side of the larger pipe—approximately 12 inches from each end. Use some type of noncombustible blocking to hold the handles in position while tack welding; then finish welding and filling until each handle is securely fastened.

The last piece to attach is the cap on one end of the large pipe. Begin by first cutting a suitably sized circle from the ⅜-inch steel plate. Then position this piece over one end of the pipe, and tack weld it in place. Finish welding around the end of the pipe to hold the cap in place. Weld carefully here as this piece acts as a hammer when pounding a post into the ground.

After all welding, clean up all joints with a grinding wheel or side-angle grinder to remove any sharp edges. Now your post pounder is ready for service.

Hay Manger

Feeding hay to horses and other livestock can become wasteful when the hay is simply put on the ground for the animals to eat. A good solution to this problem is to build a hay manger (Fig. 16.2) in the welding shop. The cost is minimal, and the hay manger will last for years. Of possible greater importance, less hay will be wasted and your horses will be healthier as a result of cleaner feed.

Begin by marking the steel drum into two equal sections. Cut the drum in half using a reciprocating saw with a metal cutting blade. While the drum could be cut in half with a cutting torch, this can be dangerous as built-up heat inside the drum could cause it to explode. For this reason, use a reciprocating saw for cutting the drum in half.

Next, place the two barrel halves together, along their sides. Measure the width and length of the two halves; then cut angle steel to make a frame that will fit over the edges of both barrel halves.

Weld the frame over the top edges of both barrel halves. The purpose of this frame is to add strength to the barrel halves and prevent livestock from being cut by the edges of the barrel halves. The frame also serves as a fastening point for the re-bar uprights.

Now turn your attention to making the legs for the manger. The leg system consists of two X-shaped brackets 18 inches long with a 6-inch extension, or foot, on each leg. An additional 2-foot-long piece of angle steel is welded to the top of each pair of legs and serves as an anchoring point for the drum halves. Build one set of legs; then build a duplicate set. Connect the two pairs of legs with re-bar crossbracing (approximately 36 inches long) welded from the top of one set of legs to the bottom of the other set of legs. Two crossbraces should be welded in place.

The next step is to weld the barrel halves to the leg assembly. Begin by placing the barrel halves (with top frame attached) on top of the leg assembly and tack weld. Then run a weld bead along the top edge of each leg assembly.

1/2" Rebar

3'

Angle iron

2'

Angle brace legs under barrels

Angle iron on edge of barrel

Top View of angle iron corner showing 45° cut

End leg & brace detail

FIGURE 16.2

Hay manger.

MATERIALS

- 40 feet of 1½-inch steel angle
- 70 feet of ½-inch re-bar
- 1 50-gallon steel drum

At this point you should have two barrel halves, framed with angle steel sitting on two pairs of legs. The legs should be securely welded to the bottom of the barrel halves. Next you can begin constructing the top half of the manger.

The top half of the manger must be large enough to accommodate a full bale of hay. Begin by cutting the re-bar into the following lengths:

16—36 inches

6—48 inches

2—18 inches

2—21 inches

Next lay out one side panel of the top half. This will consist of eight 36-inch pieces, spaced 6 inches apart and connected with three 48-inch sections. Weld the longer pieces along the top and bottom of the shorter sections, and weld the remaining 48-inch piece across the middle of the shorter pieces. Repeat for the second top panel.

Now position one top panel inside one of the barrel halves so that its location is approximately 16 inches from the center. Clamp it in place, and position the second panel so that its location is in the other barrel half, 16 inches from the center, then clamp it in place. Now on each end weld an 18-inch section of re-bar up from the bottom and one 21-inch section of re-bar between the tops. Repeat on the other end of the top sections. Last, weld the bottom of this "hay bale cage" to the inside of the drums.

After all welding, clean up any rough edges with a side-angle grinder. This will prevent horses or other livestock from being cut by sharp edges. Next you can paint the manger to prevent rust. Use a primer and paint designed for metal. Now your hay manger is ready to be pressed into service, and it should last for years.

Swing Target

Target shooting continues to gain in popularity all across America as more people take up this interesting sport. While many people practice shooting at formal indoor and outdoor ranges, many also practice shooting at informal targets out in the countryside where permitted. The home welder has a definite advantage over other shooters in that he or she can easily build targets that are unique and fun to shoot at. The design offered here in Fig. 16.3 is an adaptation of a swinging target that has been used for decades.

One word of caution when shooting at swinging targets—to avoid the danger of ricochet, never stand closer than 100 feet. Shooting at closer distances greatly increases the chance of a dangerous mishap, especially with small-caliber firearms (.22 long rifle, for example).

Begin by tracing the pattern to the steel plate. Use a soapstone pencil or welder's marking crayon for this. Next secure the steel plate with clamps to the edge of your welding bench. Then cut out the pattern with a cutting torch. The backhand cutting technique is preferred over the forehand cutting technique.

After the pattern has been cut, allow the steel plate to cool. Next, using a bench grinder or side-angle grinder, grind off any rough edges or slag. Now you must find the balance point of the target by laying it on the steel pipe, on a table or other flat surface until both the rabbit and circular target are not touching the tabletop. Then mark this location with soapstone, left to right across the pattern.

1/4" Steel Plate
1/2"galv.Pipe
Rebar

SWING TARGET

FIGURE 16.3

Swing target.

MATERIALS

- **Steel plate 24 by 8 inches, $1/4$ or $3/8$ inches thick**
- **$1/2$-inch re-bar 72 inches long**
- **$3/4$-inch galvanized steel pipe, 6 inches long**

Next you must attach the $3/4$-inch galvanized pipe across the back of the target above the balance point. By attaching the pipe at this location, the target will always remain upright. Weld the steel pipe in place.

The last step is to insert the re-bar into the steel pipe until it extends equally from both sides—roughly 33 inches from each end of the pipe. Mark these locations with soapstone. Next bend the re-bar so that it forms a 90-degree angle on both ends as it passes out of the pipe. In effect, you will be making a U shape.

Now you are ready to press your swinging target into service. To use, simply push both legs of the re-bar into the ground until the target is steady and level. When you hit the target, it will ring out, and, with larger-caliber bullets, the target will spin around.

Variations are, of course, possible with this swing target. Some suggestions include round or other shapes of targets top and bottom. Silhouettes of other animals could also be used. You could make multiple targets as well. Some shooters like to spray paint the swinging target flat black before shooting so they will know exactly where their bullets are hitting the target.

Garden Row Marker

Gardening is a wonderful pastime that can easily provide a lifetime of enjoyment. Whether you are a gardener yourself or know a gardener, you can easily make a garden row marker (Fig. 16.4) in the welding shop that will prove very useful.

Basically there are two different types of garden row markers you can make: those shaped in the form of a vegetable and square-topped markers to which self-stick labels can be attached. We will cover both types in this section.

To make a vegetable-shaped row marker, begin by tracing a pattern of a vegetable on the face of the sheet metal. Use a fine felt-tipped marker for this. If you are artistically inclined, draw the pattern free hand; if not, make a copy from a seed catalog and enlarge it to a suitable size; then transfer the pattern.

Next cut out the pattern with tin snips. Then, using a file, smooth all edges to prevent cuts when pushing the row marker into the ground.

Curled Wire

GARDEN ROW MARKERS

FIGURE 16.4

Garden row marker.

MATERIALS

For one garden row marker:
- Sheet metal: 6 by 3 inches
- Welding rod: 36 by $\frac{1}{8}$ inches

The procedure for making a square or rectangular row marker is similar. Simply cut the sheet metal into the desired sized shape—3 by 5 inches, for example—and file all four sides so there are no sharp edges.

Once the top of the marker has been cut and smoothed, the next step is to attach the welding rod. This is most easily done by tack welding the rod to the top. You will find this easier to accomplish if you flatten one end of the rod before welding. To flatten, simply give a few blows of a hammer to the end of the rod, set on top of an anvil or vise. The flattened top will be much easier to weld.

The last step is to paint the row marker. If you are making vegetable shapes, paint the top a suitable color—orange for carrots, green for peas, beet red for beets, and so on. For square-topped row markers, spray with white paint. A paint designed for metal should be used. Painting the markers will prevent rust and add a little charm to your custom-made row markers.

Metal Saw Horses

Since so many welding projects are easier to accomplish at waist level than on the ground, it will be to your advantage to build a pair of sturdy metal saw horses for the workshop. The plans given in this section (Fig. 16.5) will result in a pair of durable saw horses that will enable you to place even large projects at a comfortable working height.

Begin by cutting the ends of all legs (eight) at a 30-degree bevel. First mark the cut lines using a protractor; then make the cuts using a hack saw. Clean up the cut edges with a bench grinder or disc sander to remove any metal burrs. The fit to the top rail of the saw horse should be two well-mated surfaces. Once you have accomplished this, you can begin welding the legs in place.

—2" Pipe

METAL SAWHORSE

FIGURE 16.5

Saw horse.

MATERIALS

To make a pair of the illustrated metal saw horses, you will need the following materials:

A. 48 inches of 2-inch steel pipe, top rail (two)
B. 30 inches of 2-inch steel pipe, legs (eight)
C. 16 inches of 2-inch steel strap, braces (four)

Position one leg (B) about 6 inches in from the end of the saw horse top rail (A). Use a clamp to hold the leg in this position, and tack weld the top of the leg to the top rail. The top of the leg must be flush with the top rail. Once you have done this, you can remove the clamp and finish welding. Run a welding bead around the entire joint, using filler rod for a strong weld.

After the first leg has been attached in this manner, fasten another leg on the opposite side of the top rail. Repeat this procedure on the other end of the top rail, welding those two legs about 6 inches in from the end. Once all four legs have been attached, stand the saw horse up and inspect the work. It should stand firmly, if all welds have been made properly.

The next step is to attach the horizontal braces between both pairs of legs. Place the steel strap (C) over the inside edges of one pair of legs, and adjust it so that it is horizontal to the ground. Hold the brace in place with clamps while you tack weld it in position. Remove the clamps and run a welding bead around all mating surfaces of brace and leg. Repeat this procedure for the opposite pair of legs.

After you have welded one saw horse, weld the second horse in the same manner. You should then have a pair of metal saw horses that should last a lifetime.

Before you press your new saw horses into service, you should clean up all the welds and any sharp edges with a wire brush and disc sander. Follow this with a coat of metal primer and two coats of metal paint, if desired. This will protect the metal from rust and give a nice finish appearance.

Metal Gate

A metal gate is particularly handy around the farm or ranch, but it can also be useful in a residential setting—at the end of the driveway or as a gate for the backyard fence. The diagram (Fig. 16.6) in this section is for an 8-foot-long gate, but the design can, of course, be modified for your particular needs simply by adjusting the lengths of uprights and cross members accordingly.

This gate is attached to an upright pole with standard gate hardware. The female parts are welded on one of the side uprights, and the lag screw male parts are placed into the pole at suitable locations. For

METAL GATE

FIGURE 16.6

Gate.

long gates, such as in the illustration, three attachment points are recommended. For smaller gates—up to about 6 feet in length—two attachment points should suffice. The latch can also be standard gate hardware or, if security is a consideration, a sliding bolt that can be padlocked.

The pole that holds the gate up must be planted deeply or the gate will sag in time. A hole with a minimum depth of about 6 feet is required for this gate. Set the pole in concrete for added strength.

Begin by cutting the uprights (A) top and bottom rails (B) and middle rails (C) to the proper length with a hack saw. After cutting, clean up all the cut edges with a disc sander so that all mating surfaces are beveled slightly and form a tight-fitting butt joint. Next lay the parts out on a flat surface—the two longer rails are placed on the top and bottom of the ends, and the shorter rails are positioned at equal distances apart in the center of the gate. It is important that the gate be square. Use a metal carpenter's framing square to help you align the pieces.

Begin by tack welding the top and bottom rails (B) over the ends of the uprights (A). Check the alignment often with your framing square to be sure the basic outline of the gate remains square. After all parts have been tacked, weld each joint solidly by running a bead of welding rod all the way around these butt joints.

MATERIALS

A. 60- by 1½- by 1½-inch square steel tube, sides (two)
B. 8-foot by 1½-inch by 1½-inch square steel tube, top and bottom rails (two)
C. 7-foot, ½-inch by 1½-inch by 1½-inch square steel tube, middle rails (two)
D. Lag bolt gate hinges (three)
E. Gate latch or sliding bolt (one)

Next, reposition the middle rails (C) in the gate. Make certain that these pieces are parallel to the top and bottom rails, then tack weld them in position. The next step is to weld the middle rails in position by running a continuous bead of welding rod around each of the butt joints.

Attach the female gate hinge parts by first drilling holes at suitable locations. Insert the hinge parts, and weld them in place. If a latch is to be attached to the other upright, do that next.

After the gate has been welded, clean up all joints and edges with a disc sander so that there are no sharp edges. Next, give a coat of metal primer to all surfaces followed by two coats of exterior metal paint. You may find the priming and painting easier to accomplish if you install the gate beforehand.

Fireplace Grate

There is nothing like a blazing fire on a cold winter's night, and one way to keep the fire burning brightly is to use a fireplace grate (Fig. 16.7). A fireplace grate keeps the logs and burning materials up off the floor of the fireplace so that air can keep the fire blazing rather than smoldering. The plan given here will produce a fireplace grate measuring 16 inches deep, 12 inches high, and 24 inches wide. Check the dimensions of your fireplace and adjust the design accordingly.

Begin by measuring, marking, and cutting the steel bar. Note that one end of some parts and both ends of the other parts are cut at an angle. While all cuts can be made with a hack saw, the cutting will be much more precise if done on a metal chop saw, set to the required angle.

After all cuts have been made, place the parts into three groups: A and B, C and D, and E. Assemble and weld the first group first. Begin by laying out two A's and one B, which will form one set of legs. Weld the fitted pieces together; then make a second pair of legs in the same manner. Clean up all welds with a bench grinder or side-angle grinder to remove any rough edges.

Next, turn your attention to the second group, which are the cross parts of the

1/2" or 3/4" Bar

FIREPLACE GRATE

FIGURE 16.7

Fireplace grate.

grate: two D's and one C. Lay the pieces out, and weld one set. You will make four of these cross pieces. Clean up all welds with a bench grinder or side-angle grinder to remove any rough edges.

After all parts have been welded, it is time to assemble the grate. Begin by attaching each of three cross members (E) to the top of the leg braces (B). Attach these 3 inches in from each edge and one down the center. Use clamps to hold the cross member to the leg assemblies until tack welded; then weld the joints completely. Repeat for the second pair of legs. When finished, you will have a pair of legs holding up three bars.

Next attach the grate parts (C and D) to the cross members. Space the grate parts evenly along the cross members, and use clamps to hold in position until tack welded. Then weld the joints completely. Clean up all welds with a side-angle grinder to remove any rough edges, and the grate is ready for service.

> **≡ MATERIALS**
>
> All made from ³/₄-inch steel bar; 20 feet total:
> A. Legs 8 inches long, one end cut at a 45-degree angle
> B. Leg crossbraces (two) 16 inches long, both ends cut at a 45-degree angle
> C. Grate horizontal pieces (four) 16 inches long, both ends cut at a 30-degree angle
> D. Grate uprights (eight) four inches long, one end cut at a 30-degree angle
> E. Cross members (three) 24 inches long, ends cut square

Stationary Welding Tanks Stand

In the welding shop, tanks of oxygen and acetylene must be secured at all times to prevent mishap. While mobile carts are available, many welders prefer to secure their tanks in one location and simply run hose to the work area. A simple but effective tank stand (Fig. 16.8) can easily be built in the home welding shop for minimal cost.

Begin by cutting the 3-inch steel strap into the following lengths:

One 4 inches

One 12 inches

One 23 inches

Two 24 inches

One 27 inches

Next bend both 23- and 27-inch pieces into circles, and weld them closed. Each of these will be for the base of either an oxygen (23

I-Beam

1/4" Steel

1/4"x1 1/2" Strap

STEEL TANK STAND

Rod Holder 3" Pipe (optional)

FIGURE 16.8

Stationary welding tank stand.

MATERIALS

■ ³/₄-inch steel plate 18 by 30 inches
■ Steel strap 3 by 120 inches
■ Steel chain 24 inches long
■ Two ¹/₄-inch steel eye hooks

inches) or acetylene (27 inches) tank. Mark the location of these pieces on the ³/₄-inch steel plate, and tack weld them in place.

Now turn your attention to building the upright support, which consists of two 24-inch sections welded—edge to face—to form a T shape. On the flat side of this, attach the 4-inch steel strap at a height of about 18 inches. This will become an anchor point for the hold-down chain. Next, weld the 12-inch piece of strap on top of this column, flat side up—this arm will serve to hold oxygen and acetylene hoses when not in use.

Next, you must weld the upright post to both the base plate and the base rings. Clamp and tack weld first; then go back and make a solid welded joint. Then find the middle of the chain, and tack weld it to the inside of the upright column at a height of about 18 inches. The last step is to drill two ¹/₄-inch holes in the 4-inch section of steel strap, and install the two eye bolts, facing left and right. It may be necessary to open each of these eye bolts so that a chain link can be inserted. In use, chain is wrapped around each tank and secured in the eye bolt.

Clean up any rough edges with a side-angle grinder, and paint if desired. Be sure to use a primer and paint designed for metalwork.

Bench Grinder Stand

A bench grinder is indispensable around the home welding shop. While the ideal setup for a grinder is on a bench, this may not always be possible because of space limitations. A sturdy bench grinder stand

(Fig. 16.9) can easily be made in the home welding shop and should serve your needs well. One added benefit of this grinder stand is that it can easily be moved closer to the work area. This is a simple project that takes about one hour to complete.

Begin by finding the center of the larger of the two steel plates. Mark this location, and tack weld one end of the pipe to the plate. After the pipe has been secured with tack welding, run a weld bead around the edge of the pipe where it sits on the steel plate.

Next, locate the center of the remaining steel plate, and tack weld the other end of the steel pipe here. You will find this work easier to accomplish if you have a helper hold the upright while you weld. After tack welding, run a weld bead around the pipe where it sits on the steel plate.

Clean up any rough edges with a side-angle grinder; then turn the stand so that the larger plate sits on the floor. It is now necessary to drill four holes in the top plate for mounting your bench grinder. After the holes are drilled, mount the grinder with suitable sized bolts, and the grinder stand is ready for service.

Metal Workbench

Any workshop needs a sturdy workbench for working on projects. A workbench for an automotive or metal welding shop has to be extremely sturdy as the work you will be doing here commonly involves material of substantial weight. The metal workbench offered in this section will provide a

Pipe

GRINDER STAND

FIGURE 16.9

Bench grinder stand.

MATERIALS

- One $^3/_4$-inch steel plate, 18 by 24 inches, top
- One $^3/_4$-inch steel plate, 24 by 24 inches, bottom
- One 2-inch steel pipe 36 inches long

rugged work surface that is unaffected by hard work and should provide a lifetime of sturdy service. Additionally, this is a good beginner welding project (Fig. 16.10).

Begin by cutting all the pieces with a hack saw. Clean up all cut ends with a bench grinder so that there are no sharp metal edges. When you purchase the steel plate for the top of this workbench, have it cut at the metal fabricators and save yourself a lot of work. If this is not possible, cut the top after it is mounted on the frame.

Start welding the top frame by attaching the front and rear top supports (C) to the side supports (B). The finished frame should measure 48 by 29½ inches.

Next attach a leg (A) to each corner of the top frame. You will find this work easier to accomplish if you clamp the pieces together before

METAL WORKBENCH

FIGURE 16.10

Metal workbench.

welding. Run a continuous bead of welding rod around the edges of the mating surfaces between the tops of the legs and the top frame.

After the top and legs of the frame have been welded, attach the front rails (D) about 18 inches up from the bottom of the front legs. Install the other long rail between the rear legs in the same manner. Then attach the side support rails (E) between the side legs.

Once the basic frame has been constructed, mount the top (F). All top edges should protrude 2 inches over the top frame, all the way around the workbench. Tack or spot weld the top in place. Place these small welds about 6 inches apart on the underside, where the top sits on the top supports. Be careful not to use too much heat as this could cause the top sheet steel to warp.

After all welding has been completed, clean up all welds with a disc sander to remove any roughness. Sand the top smooth to remove any rust or rough spots as well. If desired, prime the workbench with metal primer, followed by two coats of metal paint. After the paint dries hard, this workbench is ready for use and should provide years of sturdy, dependable service.

MATERIALS

A. 28- by 2- by 2-inch steel angle, legs (four)

B. 28- by $\frac{3}{4}$-inch steel tubing, side top supports (two)

C. 48- by $\frac{3}{4}$-inch steel tubing, front and rear top supports (two)

D. 46$\frac{1}{2}$- by $\frac{3}{4}$-inch steel tubing, front and rear rails (two)

E. 24- by $\frac{3}{4}$-inch steel tubing, side rails (two)

F. 52- by 33$\frac{1}{2}$- by $\frac{3}{8}$-inch steel plate, top (one)

GLOSSARY

abrasion   A condition of wear most often caused by the rubbing together of two or more surfaces.

ac   The standard abbreviation for *alternating current.* A common type of electricity that reverses its direction of electron flow regularly and periodically.

acetylene   A gas composed of two parts carbon and two parts hydrogen. When burned in an atmosphere of oxygen, it produces one of the highest flame temperatures available.

acetylene cylinder   A specially built container used to store and ship acetylene. Occasionally called a *tank* or *bottle.*

acetylene regulator   An automatic valve used to reduce acetylene cylinder pressures to working torch pressures and to keep the flow of acetylene constant.

acid core solder   See preferred term *cored solder.*

activated rosin flux   A rosin- or resin-base flux that contains an additive to increase wetting action of solder.

alloy   A mixture of two or more metals to achieve specific qualities such as hardness and ductility.

annealing   Softening metals by heat treatment. This process most commonly involves heating the metals up to a critical temperature and then cooling them slowly.

anode   The positive terminal of an electrical circuit.

arc   The flow of electricity through a gaseous space or air gap.

arc cutting   A group of cutting processes wherein the severing or removing of metals is accomplished by melting them with the heat from an arc between an electrode and the parent metal.

arc-seam weld   A weld bead with an arc welding unit.

arc-spot weld   A spot weld made by an arc welding process.

arc voltage   The electrical potential across an arc. The pressure or voltage of an arc.

arc welding Fusing metals using the arc welding process.

automatic oxygen cutting Oxygen cutting with equipment that is fully automated, requiring only that an operator set up the work initially.

automatic welding Most commonly some type of arc welding wherein all welding operations are controlled and initiated by automation.

AWS The American Welding Society.

axis of a weld An imaginary line along the center of gravity of the weld metal and perpendicular to a cross section of the weld metal.

backfire Momentary retrogression or burning back of the torch flame into the torch tip. Immediately following the withdrawal of the tip from the work, the gases may be reignited by the hot work piece. Otherwise, the use of a lighter may be necessary.

back gouging The forming of a bevel or groove on the other side of a partially welded joint to assure penetration upon subsequent welding from that side.

backhand welding That method of welding in which the torch and rod are so disposed in the vee that the torch flame points back at the completed weld, enveloping the newly deposited metal. The rod is interposed between the torch and the weld.

balling up The formation of globules of molten brazing filler material or flux by failure to adequately wet or tin the base metal. Also a professional welder's term (slang) used to describe a job that has been done poorly.

base metal Material composing the pieces to be joined by welding. Also called *parent metal.*

bead Denotes the appearance of the finished weld and describes the neatness of the ripples formed by the metal while it was in a semiliquid state.

bevel A special preparation of metal to be welded wherein the edge is ground or cut to an angle other than 90 degrees to the surface of the parent metal.

blind joint A joint in which no portion is visible.

blowpipe An oxyacetylene torch handle.

bond Junction of the weld metal and the base metal.

braze A weld wherein coalescence is produced by heating to suitable temperatures above 800°F and by using a nonferrous filler metal having a melting point below that of the base metals. The filler metal is distributed in the joint by capillary attraction.

braze weld A weld wherein coalescence is produced by heating to a suitable temperature above 800°F by using a nonferrous filler metal having a melting point below that of the base metals. The filler metal is not distributed in the joint by capillary attraction.

Brinell test A method for determining the surface hardness of metallic materials.

bronze weld See *braze weld.*

buildup The amount of weld face or bead that extends above the surface of joined metals.

burned metal Term occasionally applied to the metal that has been combined with oxygen so that some of the carbon changed into carbon dioxide and some of the iron into iron oxide.

burning Violent combination of oxygen with any substance that produces heat. This word is sometimes used in place of the term *flame cutting.*

butt joint An assembly in which the two pieces joined are in the same plane, with the edge of one piece touching the edge of the other. See *vee groove.*

butt weld The actual weld in a butt joint.

capillary action Property of a liquid to move into small spaces if it has the ability to "wet" these surfaces.

calcium carbide CaC_2. A chemical compound of calcium and carbon usually prepared by fusing lime and coke in an electric furnace. This compound reacts with water to form acetylene gas.

carbon An element that, when combined with iron, forms various kinds of steel. In steel it is the changing carbon content that causes the physical properties of the steel. Carbon is also used in a solid form as an electrode for arc welding, as a mold to hold weld metal, and for brushes in electrical motors.

carbonizing See *carburizing.*

carburizing A carburizing flame is an oxygen-fuel gas flame with a slight excess of fuel gas.

case hardening Adding carbon to the surface of a mild-steel object and heat treating to produce a hard surface.

castings Metallic forms that are produced by pouring molten metal into a shaped container or mold.

cathode An electrical term for a *negative terminal.*

Celsius The temperature scale used in the metric system where zero represents the freezing point of water and 100 is the boiling point (at sea level). To convert to Fahrenheit, multiply by 9, divide by 5, and add 32. Celsius was the name of the Swede who invented the centigrade system. Symbol is C.

chamfering See *bevel.*

coalescence The growing together or growth into one body of the base metal parts.

coated electrode Metal rod used in arc welding that has a covering of materials to aid in the arc welding process.

complete joint penetration Joint penetration that extends completely through the joint.

concave fillet weld A fillet weld having a concave face.

concave root surface A root surface that is concave.

concave weld face A weld having the center of its face below the weld edges. An indented weld bead.

cone The conical part of a gas flame next to the orifice of the tip.

continuous weld Making the complete weld in one operation.

convex fillet weld A fillet weld having a convex face.

cored solder A solder wire or bar containing flux as a core.

corner flange weld A flange weld with only one member flanged at the location of welding.

corner joint Junction formed by the edges of two pieces of metal touching each other at an angle of about 90 degrees.

corrosive flux A flux with a residue that chemically attacks the base metal. It may be composed of inorganic salts and acid, organic salts and acids, or activated rosins and resins.

coupons Specimens cut from the weld assembly for testing purposes.

covered electrode See *coated electrode.*

cracking The action of opening a valve on a tank of fuel gas or oxygen and then closing the valve immediately.

crater A depression in the face of a weld, usually at the termination of an arc weld.

creep The gradual increase of the working pressure (as indicated on the gauge) resulting from the regulator seat not closing tightly against the inlet nozzle and permitting the high-pressure gas to leak into the low-pressure chamber. When this condition exists, the regulator should be repaired by a qualified person before use.

crown A curve or convex surface of a finished weld.

cutting attachment A device that is attached to a gas welding torch handle to convert it into an oxygen cutting torch.

cutting flame Cutting by rapid oxidation process at a high temperature. It is produced by a gas flame accompanied by a jet action that blows the oxides away from the cut.

cutting tip That part of an oxygen cutting torch from which the gases issue and burn.

cylinder A portable metallic container for storage and transmission of compressed gases.

dead annealed The result of heating a work-hardened metal to a red color and immediately quenching in water. This softens the metal and renders it workable again.

deoxidized copper Copper from which the oxygen has been removed by the addition of a deoxidizer, phosphorus, or silicon. This lowers the electrical conductivity but yields a product more suitable for oxyacetylene welding.

deposited metal Filler metal that has been added during a welding operation.

depth of fusion The distance that fusion extends into the base metal or the previous pass from the surface melted during welding.

dip brazing A brazing process in which the heat required is furnished by a molten chemical or metal bath. When a molten chemical bath is used, the bath may act as a flux. When a molten metal bath is used, the bath provides the filler metal.

dip soldering A soldering process in which the heat required is furnished by a molten metal bath that provides the solder.

direct polarity Direct current flowing from anode (base metal) to cathode (electrode). The electrode is negative, and the base metal is positive.

distortion Warping of a metal or metal surface as a result of uneven cooling.

downhand welding Welding in a flat position.

drag In oxyacetylene cutting, the amount by which the oxygen jet falls behind the perpendicular in passing through the material.

drop-thru An undesirable sagging or surface irregularity, usually encountered when brazing or welding near the solidus of the base metal. The condition is caused by overheating with rapid diffusion or alloying between the filler metal and the base metal.

ductility   The property of metals that enables them to be mechanically deformed without breaking when cold.

edge joint   A welded joint connecting the edges of two or more parallel or nearly parallel parts.

electrode   A substance that brings electricity up to the point where the arc is to be found.

elongation   The total amount of stretching of a specimen, produced in a tensile strength test.

erosion   Reducing the size of or wearing away of an object because of liquid or gas impact.

expansion   Increase in one or more of the dimensions of a body, caused usually by a rise in temperature.

explosion welding   A solid-state welding process wherein coalescence is affected by high-velocity movement produced by a controlled detonation.

face of weld   The exposed surface of a weld.

Fahrenheit   A temperature scale used in most English-speaking countries where 32°F is the temperature at which water will freeze and 212°F is the temperature at which water will boil, at sea level. Symbol is F.

ferrous metals   Those metals and alloys of which the principal base or constituent is iron. These metals are magnetic as well.

filler metal   Material to be added in making a weld.

fillet   Weld metal in the interval vertex, or corner, of the angle formed by two pieces of metal, giving the joint additional strength to withstand unusual stresses.

fillet weld   Metal fused into a corner formed by two pieces of metal whose welded surfaces are approximately 90 degrees to each other.

flame cutting   Cutting performed by an oxygen-fuel gas torch flame that has an oxygen jet.

flanged edge joint   A joint in two pieces of metal formed by flanging the edges at 90 degrees to the plates and joining with an edge weld.

flashback   The retrogression or burning back of the flame into or beyond the mixing chamber, sometimes accompanied by a hissing or squealing sound and the characteristic smoky, sharp-pointed flame of small volume. When this occurs, immediately shut off the torch oxygen valve and then the acetylene valve.

flat position   A horizontal weld on the upper side of a horizontal surface.

flowability   The ability of a molten filler metal to flow or spread over a metal surface.

flux   A chemical compound or mixture in powdered, paste, or liquid form. Its essential function is to combine with, or otherwise render harmless, those products of the welding, brazing, or soldering operation that would reduce the physical properties of the deposited metal or make the welding, brazing, or soldering operation difficult or impossible.

forehand welding   That method of welding in which the torch and rod are so disposed in the vee that the torch flame points ahead in the direction of welding and the rod precedes the torch.

fuel gases   Gases usually used with oxygen for heating such as acetylene, natural gas, propane, methoacetylene, propadyne, and other synthetic fuels and hydrocarbons.

fuse plug   A safety device employed on compressed gas cylinders consisting of a low melting point alloy, designed to relieve excessive internal pressure due to heat by melting at a predetermined temperature.

fusion   For the purposes of this book, the melting and flowing together of metals.

gas pocket   A cavity in a weld caused by entrapped fuel gas.

gas welding   A group of welding processes wherein fusion takes place as a direct result of the heat applied with a blowpipe fuel gas and oxygen. It is always best to use specific terminology when discussing welding fuels, for example, *oxyacetylene welding* and *mapp oxygen welding*.

generator   An apparatus for mechanically controlling the generation of acetylene by the reaction of calcium carbide and water.

gouging   The forming of a bevel or *groove* by removing material.

groove   The opening provided by a *grooved weld*.

grooved weld   A welding rod fused into a joint that has the base metal removed to form a V, U, or J trough at the edge of the metals being joined.

hard facing or hard surfacing   The application of a hard, wear-resistant alloy to the surface of a softer metal by an arc or gas welding process.

heat   Molecular energy in motion.

heat-affected zone   The part of the base metal that has been altered by the heat from the welding, brazing, or cutting operation but that may not have actually melted.

heat conductivity   The speed and efficiency of heat energy movement through a substance.

horizontal position   A weld performed on a horizontal seam at least partially on a vertical surface.

hose   Flexible medium used to carry gases from the regulator to the torch. It is made from rubber and reinforced with fabric.

hydrogen   A gas formed of the single element hydrogen. It is considered one of the most active gases. When combined with oxygen, it forms a very clean flame that, however, does not produce very much heat.

icicles   An undesirable condition wherein excess weld metal protrudes beyond the root of the weld.

inclusion   A gas bubble or nonmetallic particle entrapped in the weld metal as a result of improper torch flame or filler material manipulation.

incomplete fusion   Fusion that is less than complete.

inert gas   A gas that does not normally combine chemically with the base metal or filler metal.

infrared rays   Heat rays that come from both the arc and the welding flame.

inside-corner weld   Two metals fused together; one metal is held 90 degrees to the other. The fusion is performed inside the vertex of the angle.

intermittent weld   Joining two pieces and leaving unwelded sections in the joint.

joint   Where two pieces meet when a structure is made of smaller pieces.

joint design   The joint geometry together with the required dimensions of the welded joint.

joint penetration   The minimum depth a groove or flange weld extends from its face into a joint, exclusive of reinforcement.

kerf   The space from which metal has been removed by a cutting process.

keyhole   The enlarged root opening that is carried along ahead of the puddle when making an arc weld or other type of welded joint.

knee The lower arm supporting structure in a resistance-welding machine.

land The portion of the prepared edge of a part to be joined by a groove weld, which has not been beveled or grooved. Sometimes called *root face.*

lap joint A welded joint in which two overlapping parts are connected, generally by means of fillet welds.

layer A certain weld metal thickness made of one or more passes.

lens A specially treated glass through which a welder may look at an intense flame without being injured by the harmful rays or glare radiating from the flame.

liquidation The separation of a low melting constituent of an alloy from the remaining constituents, usually apparent in alloys having a wide melting range.

liquidus The lowest temperature at which a metal or an alloy is completely liquid.

low-temperature brazing That group of the brazing processes wherein the brazing alloys employed melt in the range of about 1,175 to 1,300°F and a shear (lap) joint is used.

malleable castings Cast forms of metal that have been heat-treated to reduce their brittleness.

manifold A multiple header for connection of individual gas cylinders or torch supply lines.

manual welding Welding wherein the entire welding operation is performed and controlled by hand.

mapp A stabilized methyl acetylene propadiene fuel gas often used in place of acetylene.

melting range The temperature range between solidus and liquidus.

MIG Gas metal arc welding (metal shielding gas).

mixing chamber That part of the welding blowpipe where the welding gases are intimately mixed prior to release and combustion.

multilayer welding In oxyacetylene welding, a technique in which a weld, on thick metal, is made in two or more passes.

neutral flame A flame that results from combustion of perfect proportions of oxygen and the welding gas. The most commonly used flame for oxygen-fuel gas welding.

noncorrosive flux A soldering flux that in itself and, as a residue, does not chemically attack the base metal. It is usually composed of rosin- or resin-base materials.

nonferrous   Metals containing no substantial amounts of ferrite or iron such as copper, brass, bronze, aluminum, and lead.

nozzle   See *tip.*

orifice   Opening through which gases flow. It is usually the final opening or any opening controlled by a valve.

outside-corner weld   Fusing two pieces of metal together, with the fusion taking place on the underpart of the seam.

overhead position   A weld made on the underside of the joint with the face of the weld in a horizontal plane.

overlap   Extension of the weld face metal beyond the toe of the weld.

oxidation   The process of oxygen combining with elements to form oxides.

oxide   A chemical compound resulting from the combination of oxygen and other elements.

oxidizing flame   A flame produced by an excess of oxygen in the blowpipe mixture, leaving some free oxygen that tends to burn the molten metal.

oxygen   A gas formed of the element oxygen. When it very actively supports combustion, it is called *burning*; when it slowly combines with a substance, it is called oxidation, as in rust.

oxyacetylene cutting   Cutting metal using the oxygen jet that is added to an oxygen-acetylene preheating flame.

oxyacetylene welding   A method of welding in which a fuel combination of two gases, oxygen and acetylene, is used.

oxygen cylinder   A specially built container used to store and/or transport oxygen.

oxygen hydrogen flame   The chemical combining of oxygen with the fuel gas hydrogen.

oxygen hose   See *hose.*

oxygen regulator   An automatic valve used to reduce cylinder pressures to torch pressures and to keep the working pressures constant. They are never to be used as acetylene regulators, and, in fact, the connections are different.

parent metal   See *base metal.*

pass   Weld metal created by one progression along a weld.

peening   The mechanical working of metal by means of repeated hammer blows.

penetration   The penetration of a weld is the distance from the original surface of the base metal to that point at which fusion ceases.

plug weld   Weld that holds two pieces of metal together. It is made by making a hole in one piece of metal that is then lapped over the other piece.

porosity   Presence of gas pockets or voids in the metal or weld bead.

postheating   Temperature to which a metal is heated after an operation has been performed on the metal such as welding, cutting, or forming.

preheating   The heating of a metal before an operation is performed on it such as welding, cutting, or forming.

psi   The standard abbreviation for *pounds per square inch.*

puddle   The portion of a weld that is molten at the place the heat is applied.

quench   To cool hot metal quickly by dunking in a liquid such as water or oil.

reducing flame   An oxygen fuel gas flame with a slight excess of the fuel gas.

regulator   A mechanical device for accurately controlling the pressure and flow of gases employed in welding, cutting, braze welding, and so on.

reinforcement weld   Weld metal on the face of the weld in excess of that required for the size of the weld, for the purpose of adding strength.

resistance welding   A process using the resistance of the metals being welded to the flow of electricity as the source of the heat.

reversed polarity   An electrode positive anode. Referring to dc and causing electrons to flow from the base metal to the electrode.

root of weld   That part of a weld farthest from the application of weld heat and/or filler metal side.

safety disc   A mechanical safety device designed to release at a predetermined pressure.

skull   The unmelted residue from a liquefied filler metal.

slag inclusions   Nonfused, nonmetallic substances in the weld metal.

slugging   The act of adding a separate piece or pieces of material in a joint before or during welding, resulting in a welded joint that does not comply with the original design, drawing, or specification requirements.

soldering   A means of fastening metals together by adhering another metal to the two pieces of these metals. The joining metal only is melted during the operation. The joining metal melts below 800°F.

solidus The highest temperature at which a metal or alloy is completely solid.

spatter In arc and gas welding, the metal particles expelled during the welding that do not form a part of the weld.

spelter The powdered brass used in making a typical brazed joint (lap joint).

spot weld A weld made between or upon overlapping members wherein fusion may start or occur on the faying surfaces or may have proceeded from the surface of one member. The weld cross section is approximately circular.

straight polarity An electrode negative cathode. Connecting dc to cause electrons to flow from the electrode to the base metal.

strain The reaction of an object to stress.

stress The load imposed on an object.

stress relieving Even heating of a structure to a temperature below the critical temperature followed by a slow, even cooling.

surfacing The deposition of a filler metal on a metal surface to obtain desired properties or dimensions.

sweat soldering A soldering method in which two or more parts that have been precoated with solder are reheated and assembled into a joint without the use of additional solder.

tack weld A small weld used to temporarily hold together components of an assembly until they can be welded.

tank See *cylinder.*

T joint A joint formed by placing one metal against another at an angle of 90 degrees. The edge of one metal contacts the surface of the other metal.

tensile strength Maximum pull stress, in pounds per square inch, that a specimen is capable of developing.

throat of fillet weld Distance from weld face to weld root.

TIG Tungsten inert gas welding.

tinning In soldering, the coating of the metals to be soldered with the soldering metal.

tip The part of the torch at the end where the gas exits and burns, producing the temperature flame. In resistance welding, the electrode ends are sometimes referred to as the *tip.*

toe of weld Junction of the face of the weld and the base metal.

torch The mechanism that the operator holds during gas welding and cutting, at the end of which the gases are burned to perform the various gas welding and cutting operations. Often called the *blowpipe.*

ultraviolet rays   Energy waves that emanate from the electrodes and the welding flames at a frequency in the ultraviolet ray light spectrum.

undercut   A depression at the toe of the weld that is below the surface of the base metal.

underfill   A depression on the face of the weld or root surface extending below the surface of the adjacent base metal.

vee groove   See *butt joint.*

vertical position   A type of weld where the welding is done on a vertical seam and surface.

voltage regulator   An automatic electrical control device for maintaining a constant voltage supply to the welding transformer.

welding   The art of fastening metals together by means of interfusing the metals.

welding rod   Wire that is melted into the weld metal.

welding sequence   Order in which the component parts of a structure are welded.

weldment   An assembly whose component parts are joined by the welding process.

weld metal   Fused portion of base metal or fused portion of both the base metal and the filler metal.

weld pool   The small body of molten metal created by the flame of the torch.

work hardening   The increase in strength and hardness produced by working certain metals such as iron, copper, aluminum, and nickel. It is most pronounced in cold welding.

yield strength   The stress in pounds per square inch at which a specimen assumes a specified limiting permanent set.

INDEX

ABOUT THE AUTHOR

Don Geary is the author of more than 25 how-to books covering a variety of shop and construction skills, as well as countless articles in *Popular Science, Popular Mechanics,* and *Family Handyman.* He lives with his family in Baker, Nevada.